FARMING, FOOD *and* FAMINE

Michael Witherick
with Sue Warn

Series Editor
Michael Witherick

Published in 2001 by:
Nelson Thornes Ltd
Delta Place
27 Bath Road
CHELTENHAM
GL53 7TH
United Kingdom

01 02 03 04 05 / 10 9 8 7 6 5 4 3 2 1

A catalogue record for this book is available from the British Library

ISBN 0 7487 5819 4

Page make-up and illustrations by Multiplex Techniques Ltd

Printed and bound in Great Britain by Ashford Colour Press

Acknowledgements
With thanks to the following for permission to reproduce photographs and other copyright material in this book:

Corel, Figs 3.6, 4.11, 7.3; Tesco, Fig 3.8
New Internationalist, Figs 6.3, 6.4 (from *New Internationalist*, March 1991)

Every effort has been made to contact copyright holders. The publishers apologise to anyone whose rights have been inadvertently overlooked, and will be happy to rectify any errors or omissions.

For G.B.

Contents

CHAPTER 1

Introduction

SECTION A

The scope of the book

This book is about a process, a product and a possible outcome. The process is **farming**, the product **food** and the possibility **famine**. The last is intended to signal that there are times when the supply of food can fall seriously short of meeting human needs. This unwanted scenario also reminds us that people are the vital link between these three things. People provide the labour and enterprise needed in farming. They process farm products into food, and distribute that food. They buy and consume food. They determine whether or not the supply of food is adequate by their numbers, by the productivity of their farming and by their ability to acquire food by other means (**1.1**).

Figure 1.1 The farming, food and people system

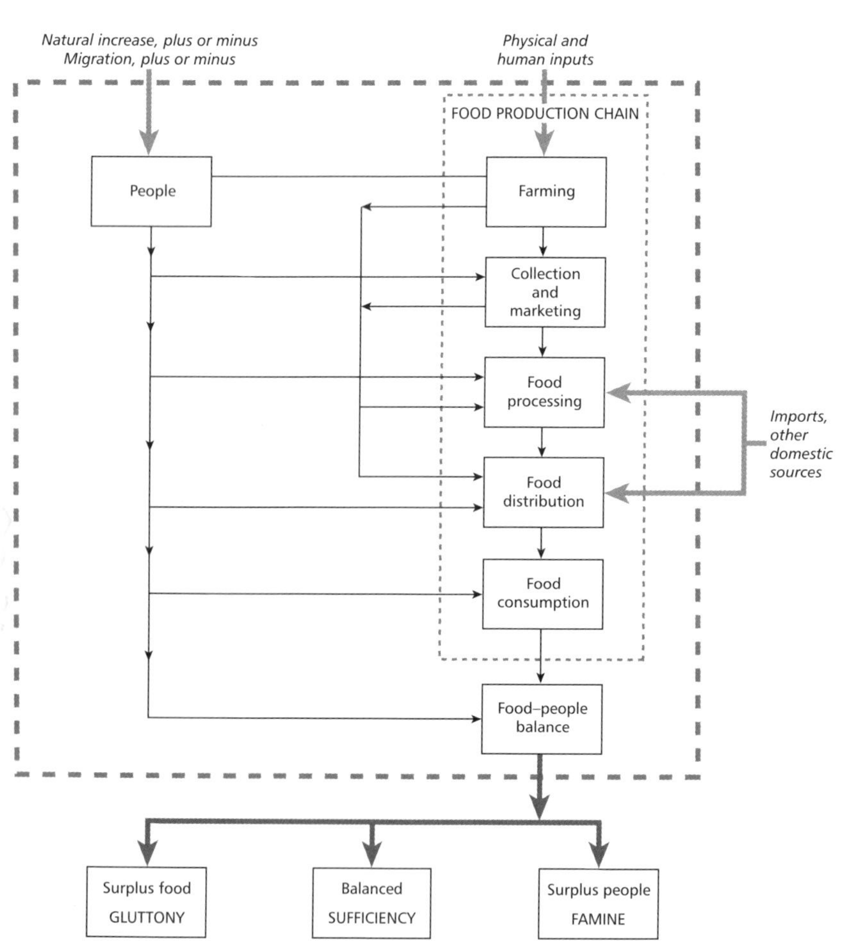

Figure **1.1** encourages us to see farming, food and people as part of one system. The inputs of the system are the normal inputs of agriculture (see **Chapters 2** and **3**) plus those other sources of food such as fishing, aquaculture and foreign trade. Farming is the starting-point of a food chain that, in more economically developed countries (MEDCs) at least, progresses through a series of stages including processing farm output into marketable food, packaging and distribution (wholesaling and retailing) and ending with consumption. The nature of the system's output is largely conditioned by what might be described, perhaps rather crudely, as the **food–people balance** (**FPB**). The nature of that balance can result in any one of three strongly contrasting situations:

- surplus food and gluttony
- balance and sufficiency
- surplus people and famine.

Review

1 What is a **system**? Does what is shown in **1.1** really qualify?

2 How might a nation acquire food by means other than agriculture?

These three scenarios are explored in **Chapter 4**. Clearly, any increase in population will alter the system's balance. So too will any raising of agricultural productivity. These changes may be expected to alter the FPB, unfavourably in the first case and favourably in the last. **Chapter 5** investigates the scope for eliminating the two extreme FPB situations and so move the world's population towards the desirable 'middle' state of 'balance and sufficiency'. That sought-after future scenario is closely tied up with ideas of **sustainability**. **Chapters 6** and **7** explore what that involves so far as food production is concerned.

SECTION B

Terms of reference

The title of this book and the previous section refer to some key terms that need to be made clear from the outset.

Farming

Does farming mean the same as thing as agriculture? Whilst few would disagree that farming is what farmers do, there is rather more scope for debating the meaning of agriculture. A liberal definition is that 'it is the science and art of cultivating the soil and rearing livestock'. A narrow definition, held by some, is that agriculture is only about cultivation and therefore excludes anything to do with livestock. Given this fundamental division of opinion, it is perhaps safer for us to turn to the term 'farming'. Even so, we need to be clear that the activities of farmers are not only to do with raising food. Forestry and the production of raw materials such as cotton and wool, tallow and hides are important lines with some farmers. Equally, in many MEDCs today, farmers are being encouraged to look outside agriculture in order to diversify their sources of income. Recreation and tourism are two obvious relative newcomers to the MEDC farm.

In this book, the two terms 'agriculture' and 'farming' will be taken as synonymous and therefore interchangeable, but preference will be given to the latter. Both are concerned, amongst other things, with the production of food – vegetable and animal – for human consumption.

Food

Given that food is absolutely basic to everyone's survival, we probably all agree that food comprises 'those substances that nourish the body'. The differences come when we examine the 'substances' consumed to do just that. Figure **1.1** has already hinted that the amount of food consumed varies from place to place, and from person to person. So too does the quality of food: its wholesomeness, dietary value and so on. The point is illustrated by **1.2**. There are also significant variations when it comes to looking in more detail at what exactly is eaten. For example, think of the eating habits of vegetarians and vegans. Think too, on a wider scale, of the impact of some of the world's major religions in determining eating habits.

Figure 1.2 Some possible indicators of hunger (1997)

	Daily per capita supply of calories	Daily per capita supply of protein (grams)	Food production per capita (1989–91 = 100)	Underweight children aged < 5 years (%)
All less economically developed countries (LEDCs)	2628	66.4	132	31
Least economically developed countries (LLEDCs)	2095	51.4	115	40
Eastern Europe and the Commonwealth of States (CIS)	2800	85.0	76	No data
MEDCs	3377	104.8	106	No data
World	2751	73.5	124	29

The degree to which agricultural products are processed before they are consumed also varies spatially. In less economically developed countries (LEDCs) a large proportion of families produce what they consume. There is little processing of that produce beyond grinding grain and cooking. The situation is totally different in MEDCs, with only a small minority producing their own food. The majority buy all or most of their food with money earned doing other things. Meeting this demand has created a huge food industry (**1.1**). Much of that industry is to do with the processing and packaging of food, from the packet of flour to the ready-made chicken korma for two. Further 'downstream' in the food production chain, there is the huge and complex distribution business that eventually delivers food to the consumer, be it via the supermarket or the local store, the restaurant or the fast-food take-away , the pub or the hotel.

There is a third component of the food industry that should not be overlooked. This is to be found at the 'upstream' end of the food chain (**1.1**). It provides the vital links between the producers and the processors or distributors. Remember that not all farm produce has to be processed before it is sold to the consumer. It is the task of these 'collectors' to scout for sources of supply and to ensure a reliable delivery of produce to the factory or outlet. Often, the networks involved in this transfer are of a truly global scale. This point also serves to underline the fact that in MEDCs, in particular, a significant proportion of the food supply is acquired from overseas. There is a huge international trade in agricultural produce and food, reflecting the heavy dependence of some countries on imported supplies.

Famine

Hunger is probably the world's most pressing problem today. We can only guess how many people there are who will go to bed hungry tonight. The highly aggregated data in **1.2** may not tell us much about the precise incidence of hunger, but the table does provide some hints. It clearly points to disturbingly strong disparities between the major global divisions. At the same time, it conceals the unpalatable truth that within each of these groupings starvation and gluttony exist side by side.

In this one paragraph and its heading, we have used several different words to describe a food shortage situation and its symptoms. Do the terms 'famine', 'hunger' and 'starvation' mean exactly the same thing?

Hunger is said to exist when a person's daily calorie intake falls below certain thresholds. For example, infants under the age of one year need to consume 800 calories a day, while an adult leading an active life needs an intake of 3600 calories. The UN Food and Agriculture Organisation (FAO) defines the border of 'under-nourishment' as a daily intake of around 1650 calories. Apply these values to the data in **1.2**, and the picture becomes rather confused. The fact of the matter is that hunger is a continuum condition, running from slight to severe. Whether or not it exists, or to what degree, depends on a number of factors that relate to age and sex, lifestyle and environment.

Estimates suggest that almost one billion people, around 15 per cent of the world's population, consume too few calories a day to support an active working life. Hunger is not necessarily life-threatening. It is debilitating and increases a person's susceptibility to all manner of disease. The point has already been made that there are degrees of hunger. It is when it becomes so acute as to lead directly to death that it is more appropriate to use the term **starvation**.

A **famine** is the outcome of prolonged periods of food shortage and hunger. It afflicts quite large areas (sometimes whole countries). It involves widespread starvation and is therefore reflected in a marked rise in death rates. It is truly a human disaster. Famines occur when crops, livestock and food supplies are destroyed by natural causes such as droughts, floods, severe winds, disease and pests. Many of the world's worst famines have occurred in arid and semi-arid areas (such as Sub-Saharan Africa) and have resulted from severe and prolonged drought. But the painful fact is that famines often have to do with human actions rather than nature. War has been a common cause. Indeed, the strategy of so many wars has been to destroy enemy crops and livestock and to cut off food supply routes. During the Second World War, the three-year siege of Leningrad (now St Petersburg) by the Germans led to one million Russians dying from starvation. Between 1967 and 1969, Nigeria cut food supply routes to Biafra (a small territory fighting for independence) and caused 1.5 million Biafrans to die as a consequence.

A fourth term, **malnutrition**, is one that commonly crops up in discussions of food and people. It literally means 'bad feeding' and is caused by an imbalance between what a person eats and what is needed to maintain good health. While it often results from eating too little, there are other causes that include an incorrect mix of protein, fat, carbohydrates, minerals and vitamins in the diet. An unbalanced diet may simply reflect the nature of the food that happens to be available in a no-choice situation. It also arises from the inability of some people to digest food properly and the failure of others to eat sensibly. So here is the warning. Whilst hunger and malnutrition are largely inseparable, malnutrition can and does prevail in areas of plenty, where people either eat to excess (perhaps here we should call it **overnutrition**?) or choose to indulge in unhealthy eating practices. So perhaps you should bear this in mind when you next tuck into some pot noodles, or refuse to eat your greens!

Review

3 Why are farmers in MEDCs such as the UK looking for alternative sources of income?

4 Find about the food taboos of at least two religions.

5 Check that you understand the difference between **hunger**, **starvation** and **famine**.

6 Explain why **malnutrition** is not something that is confined to LEDCs.

SECTION C

Food and mouths

The incidence of hunger depends very much on the FPB – the critical balance between food and mouths. In theory, the relationship is a simple one. The hope is that any increase in population is going to be more than matched by an increase in food supply. Failure to achieve this clearly spells potential danger in the form of starvation and perhaps even famine.

What sort of picture do we get if we look at what has happened to food supply and population at a global level over time? Have the two been running parallel to each other or have there been significant deviations?

Food supply

Figure 1.3 The three agricultural revolutions

The First Revolution (*circa* 10 000 BC): the birth of farming
Advances – the birth of agriculture, with the domestication of a range of crops and livestock. Cereal farming (wheat, barley, millet, rice and maize) using the plough and draught animals. Cattle, sheep, goats and pigs reared for meat and milk, as well as for wool and hides.

Outcome – farming of a sedentary, peasant kind became established.

The Second Revolution (1650–1850): the commercialisation of farming
Advances – new farming practices included crop rotation, improved varieties of root crops, better renewal of soil fertility through nitrogen-fixing plants and animal manure, and more emphasis on livestock rearing. The Industrial Revolution greatly increased the demand for food and provided the means (new modes of transport) for the movement of food to growing urban markets.

Outcome – farming became more market-oriented.

The Third Revolution (1930–present): the industrialisation of farming
Advances – four main sequential steps: farm mechanisation (the tractor, combine harvester, mechanical milkers and factory farming), chemical farming (the use of agrochemicals – pesticides, herbicides and fungicides), food manufacturing (more processing and preparation of produce before reaching the consumer) and biotechnology (genetic modification of crops, livestock and food).

Outcome – farming became big business. This phase is sometimes referred to as the Green and Gene Revolution.

The history of global agriculture may be seen as involving long periods of slow and gradual change, separated by short periods of accelerated and radical change. To date, there have been three periods of such rapid change that we might call them 'revolutions' (**1.3**). They have occurred at different times in different parts of the world. Each has spread spatially from its initial heartland. Interestingly, the First Revolution took place in what are now LEDCs. Although the Second and Third Revolutions had their birthplaces in MEDCs, they subsequently spread to other parts of the world. Some elements of the Third Revolution that have been diffused as a package to the Third World have become known as the Green Revolution, to signal the enormity of the changes and their potential impact. Contemporary but different strides made in the MEDCs are commonly referred to as the industrialisation of agriculture or the Gene Revolution.

Each of these revolutions may be seen as prompting a hike in food production, moving global food supply to sustained higher levels.

Population

But what has happened to the world's population during the span of the three agricultural revolutions? The answer is that it has been growing at an ever-increasing rate. The growth curve, particularly since the 18th century, has been an exponential one. Global population has recently passed the 6 billion mark. But there is one recent glimmer of encouragement. At the very end of the 20th century, the **rate** of population growth began to slacken for the first time. But be clear, global population is still growing. Current estimates reckon that global numbers will not peak before 2050, by which time there will be 8 billion people.

The balance

The nature of the FPB during this time is much less clear-cut. Basically, there are fundamentally different views on this whole issue of the relationship between population numbers and food supply.

Perhaps the debate began with Thomas Malthus, writing at the time of the Second Agricultural Revolution. He based his theory on two principles:

- that in the absence of any checks, population has the potential to grow at a geometric rate – in other words, it could double roughly every 25 years
- that at best agricultural production increases at an arithmetic rate.

Given the nature of the two growth rates, population growth may be expected gradually to outstrip any increase in food supply (**1.4**). In Malthus' view there was clearly a 'ceiling' to population growth set by food supply. He suggested that as that ceiling was approached, population growth would be curbed by one or both of two mechanisms:

- **preventive checks** – reducing fertility (delayed marriage, celibacy and so on)
- **positive checks** – increasing mortality (malnutrition, starvation, war and infanticide).

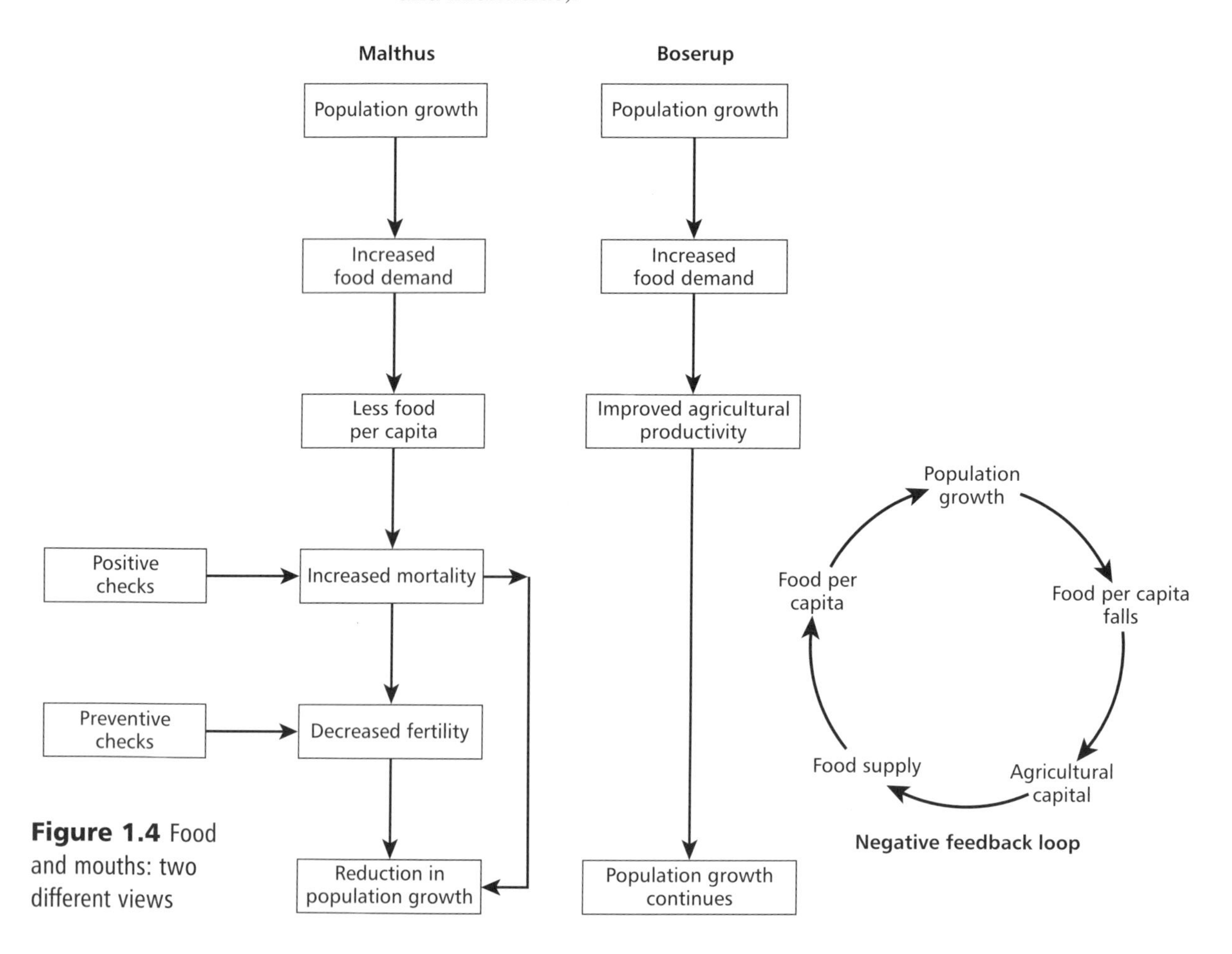

Figure 1.4 Food and mouths: two different views

In 1965 Ester Boserup, a Danish economist, put forward a view that was diametrically opposed to that of Malthus (**1.4**). Her view centred on the idea that 'agricultural developments are caused by population trends, rather than the other way round'. Certainly, there is evidence to support her claim that if food production per capita falls to a value desired by people, there will be a tendency to increase investment in food production. By this negative feedback, food per capita increases, and so too population (**1.4**). There are clear signs from the Third World of an gradual intensification of farming. So Boserup's view is a very much more optimistic view of the world and of the FPB. Some would say that it is too optimistic and that it ducks the key issue of whether or not there will come a time when the food supply can no longer be increased.

A little later, a report entitled *The Limits to Growth* (published by the Club of Rome in 1972) countered with an altogether gloomier view of the global situation. On the basis of projecting trends observed between 1900 and 1970, the report concluded that the limits to human population growth on this planet would be reached before the end of the 21st century. In a sense, the report was underlining Malthus' basic point that population growth is the cause rather than the symptom of any imbalance between food and mouths.

Supporters of what is called the **neo-Malthusian view** stress that the human race now has the ability to control its numbers in a peaceful way by means of contraceptives and other forms of fertility control. That was not really the case 200 years ago! The lowering of rates of fertility and population growth in MEDCs is partly explained by this particular 'preventive check'. But you will be quick to point out that it does not seem to have done much as yet for the LEDCs. Clearly, there are difficulties here related to all sorts of issues – traditional beliefs, education, as well as access to contraception devices. But, as pointed out earlier, there is this one ray of hope. Rates of population growth in the Third World have recently started to fall. The fall seems to be 'crisis-led', with people choosing to have smaller families when faced by increased hardship, particularly food shortages (see **Chapter 7**).

Review

7 How true would it be to say that the advances of the Third Agricultural Revolution promise few benefits for LEDCs?

8 Which do you think is more readily attainable – decreased fertility or increased food production? Give your reasons.

SECTION D

The nature and role of farming

As an economic activity, farming is unusual in that it:

- depends on biological processes
- requires large amounts of land
- is strongly influenced by the physical environment
- is immobile in terms of location

- is largely undertaken in small production units
- involves individual rather than corporate decision-making.

Like other industries, farming is valued nationally because its can provide exports and reduce imports. It also offers something that is deemed to be important by many nations – food security.

Although the main objective of farming is to produce food, it does have other purposes (**1.5**). Not only that, but the precise role that it plays in modern society depends on where you are on the globe. There are broad differences between the North (the wealthier nations of North America, Europe, the former Soviet Union, Japan, Australia and New Zealand) and the South (the less developed countries of the Tropics and Sub-tropics – often regarded as being synonymous with the Third World) and between MEDCs and LEDCs.

Figure 1.5 The purposes of agriculture

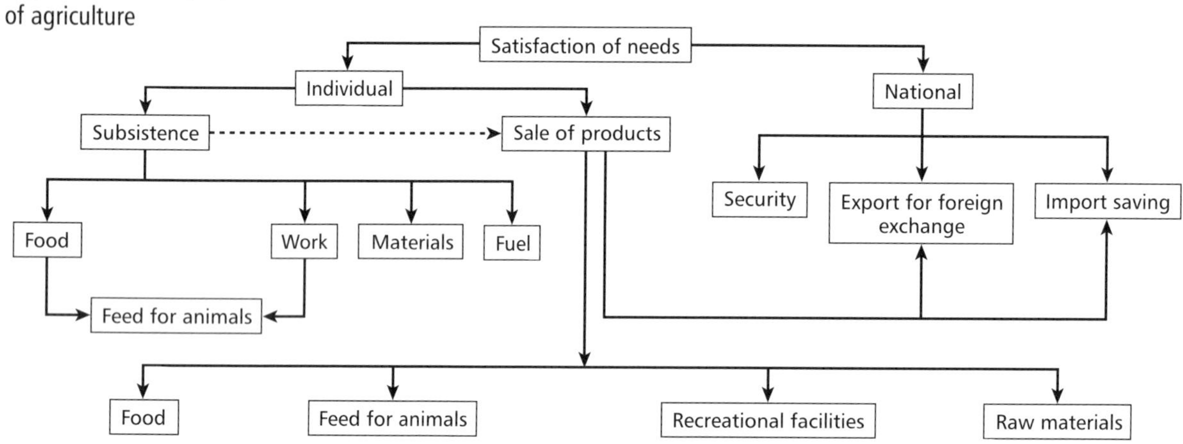

In the LEDCs of the South, agriculture is the major land use and the major employer. Most of the food produced is for the consumption of the producing family. Any surpluses of food, along with the output of any sidelines (hides, fuelwood and so on) may be traded for other commodities. It is by far the largest source of work, but because it is mainly subsistence in character, the contribution to GNP is not great (generally around 25 per cent).

In the MEDCs of the North, agriculture is also the largest single land use, but is insignificant in terms of employment (generally less than 10 per cent of total employed) and its contribution to GDP (generally around 5 per cent). The food that is produced is mainly for sale either to food processors or directly to retail outlets and consumers. Farmers have added other strings to their bows in order to offset declining demand for and falling prices of agricultural products.

Figure 1.6 The agricultural curve

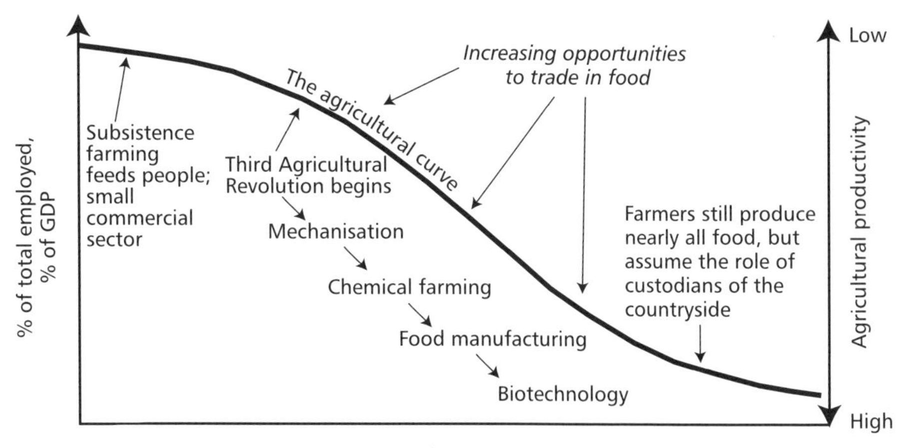

These two stereotypes of the LEDC and MEDC scenarios allow us to imagine an S-shaped agricultural curve, which largely reflects the impact of the Third Agricultural Revolution (**1.6**). The curve shows a strange paradox, namely that as agriculture's importance in employment and GDP terms falls away during the course of economic development, its productivity and ability to feed increase. At the very end of the curve – the situation today – the state of agriculture in MEDCs is such that farmers are becoming less producers of food and more general custodians of the countryside.

Review

9 What do you think is meant by **food security**? Why might it be seen as being important?

10 Choose two countries (one an LEDC and the other an MEDC) and collect data to show differences in the character and importance of agriculture.

11 Divide the agricultural curve in **1.6** into a number of stages and describe the essential features of each of those stages.

SECTION E

Topical issues

The content of this book is highly topical. Some of that content is contentious, in that touches on a whole range of issues that are very much in the limelight thanks to media attention. Figure **1.7** indicates most of those that are examined in the following chapters and attempts to classify them, perhaps not entirely satisfactorily, under one of four headings.

Classification	Issue	Where discussed
Environmental	Farming beyond the limits The agricultural landscape Impacts of modern high-tech farming Irrigation Organic farming The risks of genetically modifed (GM) crops	Chapter 2 Chapter 2 Chapter 3 Chapters 5 and 7 Chapters 5 and 7 Chapter 6
Economic	The pricing of food The power of transnational corporations (TNCs) Subsistence versus commerce Free trade Poor farmers Sustainability	Chapter 3 Chapters 3 and 6 Chapter 4 Chapter 5 Chapter 5 Chapter 7
Political	TNCs and control of the food chain Government intervention Who rules the countryside? Land reform Trade and aid Food security	Chapters 3 and 6 Chapter 3 Chapter 5 Chapter 5 Chapters 5 and 6 Chapters 4–7
Social	Health Helping the hungry Animal welfare Vegetarianism Population control Food quality and safety	Chapters 4 and 7 Chapters 5 and 7 Chapter 6 Chapter 6 Chapter 7 Chapter 6

Figure 1.7 Some issues of farming, food and famine

Review

12 Are you able to 'flesh out' any of the issues in **1.7** before reading the relevant chapter?

13 Remember to return to this when you have read the book:

- Do you think that the issues have been correctly classified?
- Have any been omitted from the table and/or the book?

Enquiry

Investigate the causes and dimensions of one recent major famine.

CHAPTER 2

Farming as agro-ecosystems

SECTION A

Ecosystems and agro-ecosystems

One of the distinguishing features of farming mentioned in the previous chapter is its exploitation of biological processes. In the natural world, biological processes are an integral part of ecosystems and involve plants, animals and micro-organisms. In agriculture, ecosystems are so modified by human actions that they become systems of a rather different kind, commonly referred to as **agro-ecosystems** (**2.1**).

Figure 2.1 The agro-ecosystem

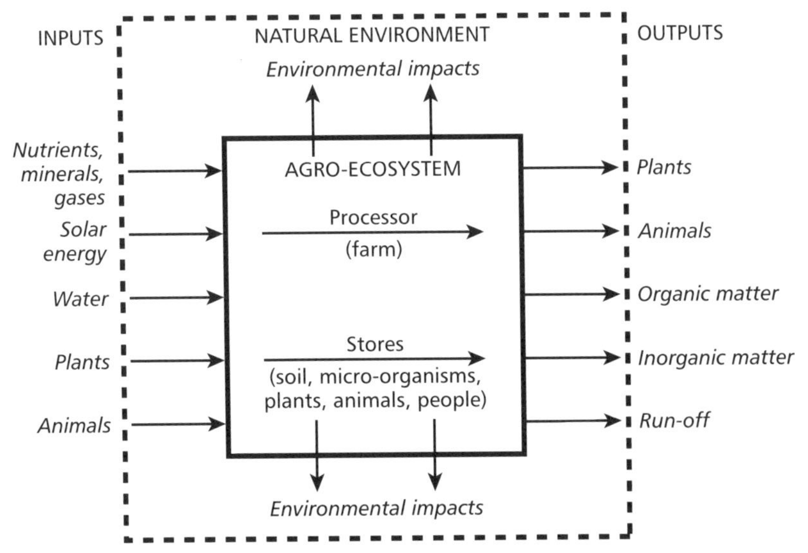

An open system has its inputs and outputs, but at its heart there are the processes (transfers or throughputs) and the stores (**2.1**). In the case of the agro-ecosystem, the farm may be seen as both the **processor** and the **store**. In this case, the store may be thought of as comprising the soil, plants, micro-organisms, animals and people. Another important difference between a natural ecosystem and an agro-ecosystem is the simplification of the latter to a relatively few species of crop and animal (**2.2**). There is a significant reduction of biodiversity. In general, the greater the degree of simplification, the greater is the amount of energy that must be introduced into the system and the greater is its potential instability. The simplification also applies to the food chain and trophic levels, the latter being reduced to two, or at most three.

Another noteworthy difference is the much more 'open' nature of the agro-ecosystem (**2.2**). On the input side, the human hand is to be seen in a number of different forms:

- moving species to new environments and therefore changing the character of plant inputs in particular
- meeting the natural resource needs of plants and animals through manipulation of the natural environment – by irrigation, the application of fertilisers and so on

Natural ecosystems	Agro-ecosystems
A rich diversity of plants and animals	Little diversity. Tendency to monoculture of both crops and livestock
A complex structure, involving levels and food chains	A simple structure with short food chains – animals account for a larger proportion of the biomass
All dead organic matter is recycled	Very little energy finds its way into dead and decaying matter in the soil
Nutrients are continually recycled, and rather slowly – there is little input of organic matter from outside	Nutrient cycling speeded up by inputs of fertilisers, but harvesting removes nutrients from the system
There is very limited movement of energy and materials across ecosystem boundaries	There are sizeable external exchanges of energy and materials

Figure 2.2 Differences between natural and agro-ecosystems

- improving the growth performances of plants and animals – by selective breeding and genetic engineering, and by the application of pesticides and fungicides.

On the output side, the harvesting of crops and livestock is the most obvious human interference, causing a substantial 'leakage' of energy and material from the agro-ecosystem.

Thus agro-ecosystems may be seen as involving flows of energy and nutrients. It will be shown later in this chapter that they can display considerable diversity, particularly with respect to their productivity and output (**Section D**). Agro-ecosystems also differ in terms of their impact on the environment (**Section C**) and their degree of sustainability (see **Chapter 7 Section E**).

Review

1 Define the term **agro-ecosystem**.

2 What is it that causes an agro-ecosystem to 'diverge' from its original ecosystem?

3 Why is an agro-ecosystem a more open system than an ecosystem?

SECTION B

Environmental factors and limits to growth

Despite all the progress of the last 50 years or so, agriculture even in the MEDCs is still dependent on the resources of the natural environment. Temperature, rainfall, soil and drainage are among the more influential physical conditions, and in general terms crops are more 'sensitive' than livestock. Although farmers can modify environmental conditions, they can do so only to a limited degree. Furthermore, the more that is done in this direction, they more it costs. So, sooner or later, there comes a point at which striving to change physical circumstances ceases to be economic. In short, the distribution of any agricultural activity is still ultimately controlled by its particular physical needs (**2.3**).

Although it is possible to create artificial climatic conditions in greenhouses and poly-tunnels, the costs are high. For this reason, climate remains a powerful influence on the distribution patterns of crops, and to a lesser extent on the distributions of livestock. In the latter, barns, stables and all manner of roofed structures are able to provide some measure of protection from the climatic elements.

Solar radiation

At the root of all plant growth is solar radiation and the process of **photosynthesis**, whereby plants transform solar energy into chemical energy (**2.3**). It is this chemical energy that fuels plant growth. The amount of solar radiation received at the Earth's surface is a function of:

- latitude
- time of year
- day length
- cloudiness.

There are, therefore, significant variations in the availability and intensity of sunlight and, in turn, in the capacity of plants to turn solar into chemical energy and then that into growth. Generally speaking, the highest receipts of solar radiation occur in the arid Tropics and Sub-tropics where cloud cover is least.

Figure 2.3 Environmental factors affecting crop growth

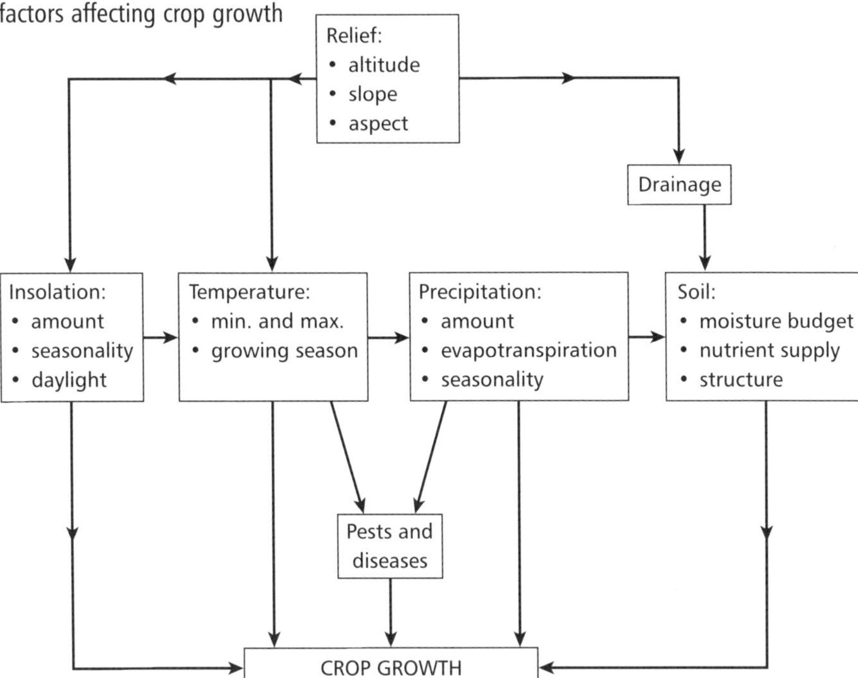

Temperature

Temperature affects the distribution of crops in a number of ways (**2.3**). All crops have a minimum temperature requirement, below which they cannot grow. Equally, many crops have a maximum temperature threshold, beyond which they cannot survive. The majority of cereals in temperate regions will germinate and develop at average temperatures as low as 5°C and as high as 37°C. These minimum and maximum temperatures help to define what is known as the **thermal growing season**. This is the number of days a year when the mean temperature is above a prescribed minimum. In middle and high latitudes, this is often taken to be a soil temperature of 6°C. Using this particular value, it is not until the latitude of southern Spain that the thermal growing season reaches 365 days. Within England and Wales, it ranges from just over 300 days in South-west England to less than 200 days in the Northern Pennines and the uplands of Mid- and North Wales.

Precipitation

Equatorwards of latitude 35 degrees, a thermal growing season of 365 days is more or less guaranteed. Here it is precipitation and soil moisture that set limits on crop and livestock production (**2.3**). There is a **hydrological growing season**, defined as the number of days on which there is sufficient soil moisture for crop growth. This in turn is conditioned not so much by how much rain falls, but by the effectiveness of that which falls. **Precipitation effectiveness** is the amount of precipitation available to plants after losses due to evaporation and transpiration (often referred to as **evapotranspiration**) have been taken into account. Thus temperatures are an important factor in the equation because they have a direct bearing on the rate of evapotranspiration. Broadly speaking, the higher the temperature, the higher is the rate.

There is another aspect of precipitation that impacts on the hydrological growing season (**2.3**) – its **seasonality**. Does most of the annual precipitation fall in one short season or does it occur fairly frequently throughout the year? The latter circumstance is likely to yield a longer hydrological growing season. How the seasonality of precipitation relates to that of temperatures is also important. Does the bulk of the precipitation fall in winter or in summer? If the former, then losses due to evaporation and transpiration will be relatively low. While losses will be higher in summer, most crops require good amounts of moisture during the thermal growing season. The intensity and form of precipitation are two more factors that have a bearing on the length of the hydrological growing season.

Relief

There are three characteristics that need to be considered – altitude, aspect and angle of slope. Their impacts on crop growth tend to be indirect (**2.3**). For example, the effects of **altitude** are felt through both temperature and precipitation. On average, temperatures fall by 0.6°C every 100 metres. There is both good and the bad news here. Within the Tropics, temperatures that otherwise would have exceeded the maximum for particular crops can be lowered to more agreeable levels. Conversely, in higher latitudes, the lowering of temperatures can be such as to make the thermal growing season too short.

Generally speaking, precipitation increases with altitude. This, together with the impact of lower temperatures, can result in precipitation falling as snow rather than rain. **Aspect** takes on increasing significance with rising altitude. Think of the differences in precipitation between windward and leeward slopes, and in temperature between north- and south-facing slopes in both hemispheres.

The gradient, or **slope angle**, is significant in agriculture in two respects. Steep slopes are vulnerable to soil erosion, particularly if they are rather bare of vegetation or are ploughed. The widespread terracing of slopes, especially in Asia, is one way of getting round this problem. Steep slopes also pose a challenge in that where they are greater than 15 degrees it is not safe to use machinery such as tractors. Without machinery, ploughing and harvesting become difficult; with it, they become hazardous. For these reasons, farmers do not normally grow crops on slopes steeper than 10 degrees.

One final impact of slopes is on drainage conditions. The steeper the slope, the greater will be the proportion of precipitation removed by surface run-off; less moisture will enter the ground. Where land is level, notably more precipitation will enter the ground by means of infiltration; moisture levels in the ground will tend to remain relatively high. These relationships link up with the next feature of the natural environment that has a direct bearing on agriculture.

Soil

Of all the properties of a soil, two are particularly important in the context of agriculture – soil moisture and nutrient supply (**2.3**). So far as the first is concerned, **2.4** shows that **soil moisture budgets** typically involve three different states:

- water surplus
- water deficit
- recharge.

It is the changing balance of precipitation on the one hand and evapotranspiration on the other that determine the nature of the annual cycle. A critical phase in that cycle is known as **field capacity**. This state is achieved when all the gravity water in a soil has drained away (usually several days to even weeks) after the cessation of rainfall. The remaining capillary moisture is sufficient to provide the needs of growing plants until such time as evapotranspiration reduces it to levels well below the field capacity. The length of the field capacity period is of great importance to farmers. As soon as the soil regains its maximum moisture, it becomes saturated and it is difficult to work with machinery without causing permanent damage. Likewise, the length of the period when soil moisture is below field capacity (when there is a moisture deficit) is critical. Soon after its onset, plant wilt will occur unless there is a resort to irrigation.

Figure 2.4 A typical soil moisture budget in England

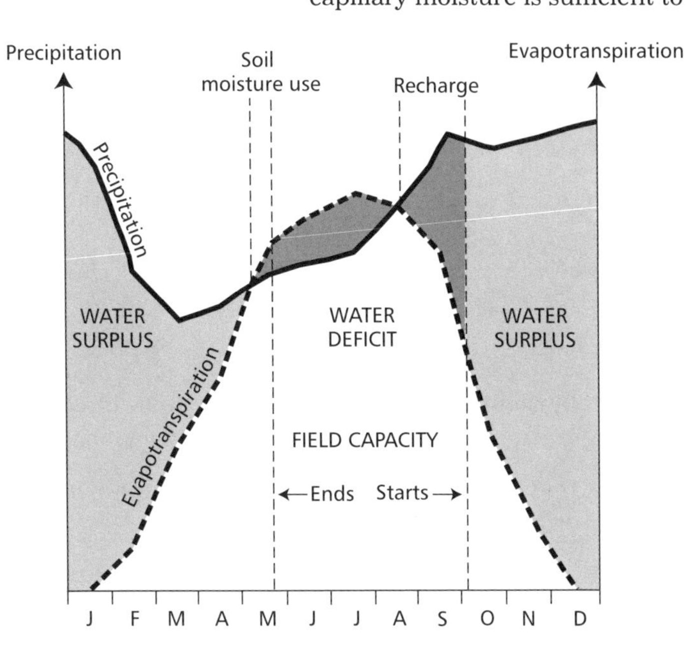

The soil is vital to plant growth through its supply of some 16 chemical elements, of which nitrogen, phosphorous and potassium are the most important. The supply is partly conditioned by parent material, while retention of the **nutrient supply** depends on soil texture and structure. Clay soils, for example, tend to retain a larger nutrient supply than sandy soils, from which nutrients are readily leached.

Soils are perhaps one environmental resource that has declined in significance as a factor affecting agriculture. This is because soils can now be improved or modified by deep ploughing, draining, liming and fertilising. The costs of doing this are not necessarily prohibitive.

Pests and disease

There have been times in the past when pests and disease have had dramatic negative effects on agriculture (**2.3**). The Irish famine of the 1840s was caused by the outbreak of a fungal disease known as potato blight. The disease known as phylloxera wiped out large areas of wine production in southern Europe during the late 19th century. Locust swarms plagued the Middle East during biblical times, leaving decimated crops in their wake. Until recently, the tsetse fly was a pest of such magnitude that it discouraged livestock rearing in low-lying tropical regions. One of the benefits of modern advances in science has been the virtual eradication of many previously damaging pests and diseases. However, there have been unwanted side-effects, as for example when toxic chemicals used in fungicides and pesticides have entered the food chain. Every now and again, there are reminders that the battle against pests and diseases has not been completely won. At the time of writing (March 2001), the UK is suffering a devastating outbreak of 'foot and mouth' disease, the first for nearly 40 years.

Review

4 How true is it to say that in the UK the thermal growing season is more significant than the hydrological growing season?

5 Explain:
- what is meant by **field capacity**
- why it is important in the growing of crops.

6 Produce a version of **2.4** to show the main features of the soil moisture cycle:
- close to the Equator
- in central Chile.

Explain what you have done.

7 Can you think of any diseases and pests that continue to threaten agriculture?

8 Why do you think that crops are more affected by environmental conditions than livestock?

9 Of all the environmental factors affecting agriculture, which do you think is the most important? Give your reasons.

Environmental impacts

One aspect of farming that tends to be overlooked when adopting an agro-ecosystems view is its environmental impacts. Here, possibly, is a limitation of the systems approach. It tends to see each system (no matter of what kind) as being rather 'isolated' from its surroundings. Those surroundings are acknowledged simply as a source of inputs and a destination for outputs. So how do we represent those impacts in the systems diagram? The question may be resolved in one of two ways. Either those impacts are recognised as outputs, or they are seen as a distinct brand of outcome along the interface between the system and its surroundings. The latter option has been followed in **2.1**, while **2.5** details the more notable impacts.

Landscape heritage

It would be true to say that virtually all types of agriculture mark and change the environment in some way or another. Indeed, agriculture generally has created its own distinctive landscapes. This is hardly surprising, bearing in mind that agriculture is the world's largest single user of land and one of the oldest economic activities.

So long is the history of agriculture and so widespread has been its impress on the landscape that we may regard the agricultural landscape and the cultural landscape as one and the same thing. In so many parts of the world, the agricultural landscape is an integral part of a region's heritage. Just look at a rural area near your home. What you see might include:

- a patchwork of irregularly shaped fields that has not changed much in its general configuration in 200 years
- fields defined by hedgerows or stone walls of considerable age
- fields showing the ridge-and-furrow marks of the old open-field system and of early attempts to improve drainage
- a fairly regular network of farmhouses and farm buildings, some of which may well date back to the Middle Ages.

But, of course, the precise character of the agricultural landscape varies enormously from region to region, country to country and continent to continent. That character is also changing, subtly and slowly in some locations, dramatically in others.

Up until the 20th century, the evolution of the agricultural landscape was a fairly benign form of environmental change. That has now altered quite dramatically. The

Figure 2.5 Some environmental impacts of agriculture

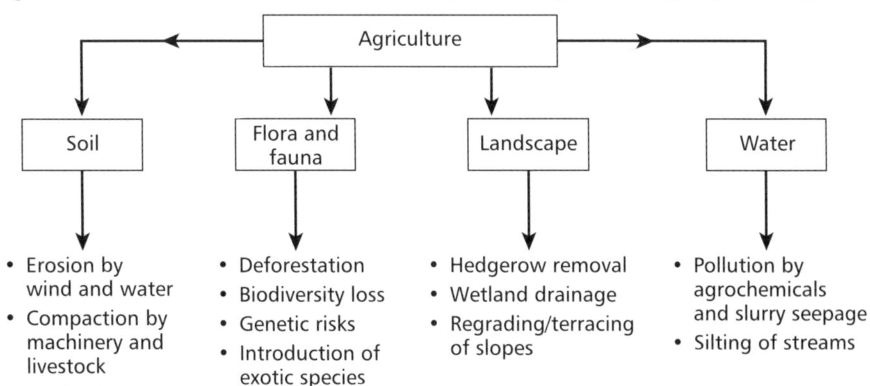

villain of the piece in MEDCs is technological progress, while in LEDCs it is population growth. In the former, it is the damage inflicted by modern 'industrialised' farming, while in the latter the two main environmental costs are deforestation and soil erosion.

The scars of modern farming

The environmental effects of modern farming are the outcome of manipulating agro-ecosystems in two ways (**2.5**):

- **Increasing energy flows** within the system by the deliberate and increasing application of fertilisers, pesticides, fungicides and herbicides. The leaching of these agro-chemicals and run-off from the storage of animal manure lead to pollution, particularly of water resources. Mechanisation is something else aimed at increasing energy flows. Its environmental costs include compaction of the soil, pollution by fuel spillages and consumption of nonrenewable resources.
- **Modifying the 'natural' components** of the system. This embraces a whole range of things, from removing hedgerows in order to enlarge fields to draining wetlands, from felling woodlands to ploughing up moorland. All this is done to increase agricultural space and productivity. Under this heading, there is also the highly topical matter of genetic engineering. The issues of cloning livestock and genetically modifying crops are discussed in **Chapter 6**.

Soil erosion

This occurs when the surface layer of productive topsoil is removed by either wind or water (**2.5**). Whilst it is a 'natural' process, encouraged by particular combinations of climate, relief, soil and vegetation, there is no doubt that it is made infinitely worse by people. This is particularly so when the natural vegetation cover is removed as a prelude to farming (**2.6**).

Figure 2.6 Soil erosion following deforestation

Soil lost to wind and water ranges from 5–10 tonnes per hectare annually in Africa, Europe and Australia to 10–20 tonnes in the Americas and nearly 30 tonnes in Asia. About half of the world's cropland is so badly managed that it is losing topsoil at the rate of 7 per cent per decade. Soil fertility is being seriously reduced, with possible catastrophic consequences in the not-so-distant future.

Wind erosion is one of the key components of desertification. It is most severe in arid and semi-arid areas exposed to overgrazing by livestock. Over 20 per cent of Africa north of the Equator is susceptible to soil erosion; in the Middle East the figure is 35 per cent. Between them, Africa and Asia lose almost a billion tonnes of soil annually just through wind erosion. In extreme conditions, up to 150 tonnes of soil can be blown from a hectare of land in an hour.

Soil erosion by water occurs where steep slopes are farmed or where slopes are left exposed for any length of time. It is estimated that 25 billion tonnes of soil are lost to water erosion each year. Two of the world's largest rivers, the Yangtze in China and the Ganges in India, transport around 3 billion tonnes of soil each year. Soil is washed into rivers, where it can silt up river channels and lakes and increase the incidence of flooding. Where it is washed into coastal waters, it can seriously damage fishing grounds.

It is important to remember that soil erosion is not confined to LEDCs. The great 'dust bowls' of the American Mid-West during the 1930s were the result of wind erosion and huge areas of once productive farmland were lost. This disaster was caused by careless cultivation and overgrazing. Soil erosion (both wind and water) is a problem even in the UK, being exacerbated by:

- the decline in organic matter content of soils (an outcome of the use of inorganic fertilisers and other agro-chemicals)
- the compaction of soil by machinery and livestock
- the exposure of arable land between harvesting and sowing.

Salinisation is another form of soil degradation. It involves the concentration of sodium, potassium and magnesium salts in the soil. This leads to the capillary rise of salt solutions from groundwater and the development of salt crystals, either in the upper layers of the soil or as a surface crust. It also results from excessive irrigation in hot, dry climates. The resulting salt pan is poisonous to agricultural crops.

Deforestation

The clearance of forest to create agricultural land is a major aggravator of soil erosion. It is also an adverse major environmental impact in its own right, often leading to a serious loss of biodiversity and certainly contributing to global warming (**2.5**). Globally, there 21.1 million km^2 of temperate forest and 12 million km^2 of tropical forest. Nearly 17 million hectares of tropical forest disappear each year as a result of deforestation – that is, about 1 per cent of the total. This is replaced by cropland or pasture, but remember that it is not only the need for more farmland that lies behind this change. Much tropical forest is being 'cropped' (once and for all) for its valuable timber. In temperate regions, primary forest growth has been largely replaced by managed plantations that are less valuable for wildlife.

Finally, it is worth pointing out that agro-ecosystems have themselves been the victims of pollution. They have been unwilling recipients of unwanted inputs from the surrounding non-agricultural environment. These have led to the pollution of the air, soil and water entering the agro-ecosystem. Agriculture is also increasingly threatened by urbanisation, greenfield industrialisation and transport systems, as well as by the demands of leisure and tourism. These pressures are particularly strong in the rural–urban fringe.

Case study: Environmental refugees

Environmental refugees are people who can no longer gain a secure livelihood in their homelands because of drought, soil erosion, desertification and other environmental problems. They are the human face of the environmental impacts just outlined, for most of these problems have a link, directly or indirectly, with agriculture and food supply; they are aggravated by the pressures of population growth and poverty.

It is estimated that there are well over 25 million environmental refugees today, or nearly 0.5 per cent of the world's population – one person in every 200. The total figure compares with 18 million officially recognised 'refugees' – people who are forced to move for political, religious or ethnic reasons. Much of the refugee movement takes place within rather than across national frontiers. For this reason, it does not hit the headlines very often, and this in turn may mean that it is seriously under-estimated.

If present trends continue, the number of environmental refugees could well reach 200 million by 2050 (possibly 2 per cent of global population). In short, environmental refugees are fast becoming one of the foremost human crises of our time. The costs are enormous. In addition to the trauma, distress and alienation acutely felt by individual migrants, there are more general costs. These include:

- a reception in the destination area that is often far from welcoming
- tensions between the customs, dietary preferences and religions of the refugees and those of the host community
- the overloading of services.

Given that the main 'push' factors are food insecurity and hunger (if not starvation), the solution to the crisis is more likely to lie in:

- finding modes of food production that are environmentally sustainable, even in these marginal areas
- spreading the necessary know-how
- putting the brake on population growth.

Such pre-emptive actions, which tackle the sources of the problem, are a better longer-term bet than directing scarce resources to setting up refugee camps and embarking on resettlement programmes that rarely succeed.

Review

10 Is modern agriculture all bad news for the environment?

11 Explain the links between soil erosion and deforestation.

12 Suggest reasons why the environmental impacts of agriculture vary from place to place.

13 Are you able to think of any other environmental impacts of agriculture that have not been covered in this section? Write brief notes about each of your additional impacts.

Types of agriculture

Given the huge variety of environmental conditions encountered on Earth, together with the immense differences in levels of development that separate societies around the world, it is hardly surprising that global agriculture should embrace an enormous diversity of types. A number of different criteria may be used in seeking to classify agricultural systems; seven have been used in **2.7**. No doubt there are more that could be used. When applying each of these criteria, it is tempting to define a set of watertight types. In reality, there is in most cases a continuum or gradation from one 'pole' or extreme to the other. With input, for example, the transition from extensive to intensive is smooth rather than stepped. There is an almost infinite number of different states between the two 'poles'. Similarly, with government, it is recognised that the degree of government intervention in agriculture runs from minimal in the completely capitalist economy (the USA) through to total in the comprehensively socialist economy (North Korea).

Another point to remember is that these criteria are not mutually exclusive. Any one agricultural system may be checked off against all seven criteria. For example, a cereal farm in East Anglia may be classified as temperate lowland, sedentary, owner-occupied, intensive, arable, capitalist and commercial. In contrast, the agricultural practices of the Masai people in East Africa may be classified as tropical upland, shifting, communal, extensive, pastoral and subsistence.

Looking at **2.7**, you might be tempted to ask: Which of the seven criteria is the most telling and useful? The answer may sound an evasive one, but importance is relative and so too is value – much depends on the particular view being taken of agriculture.

Figure 2.7 Classifying agricultural systems

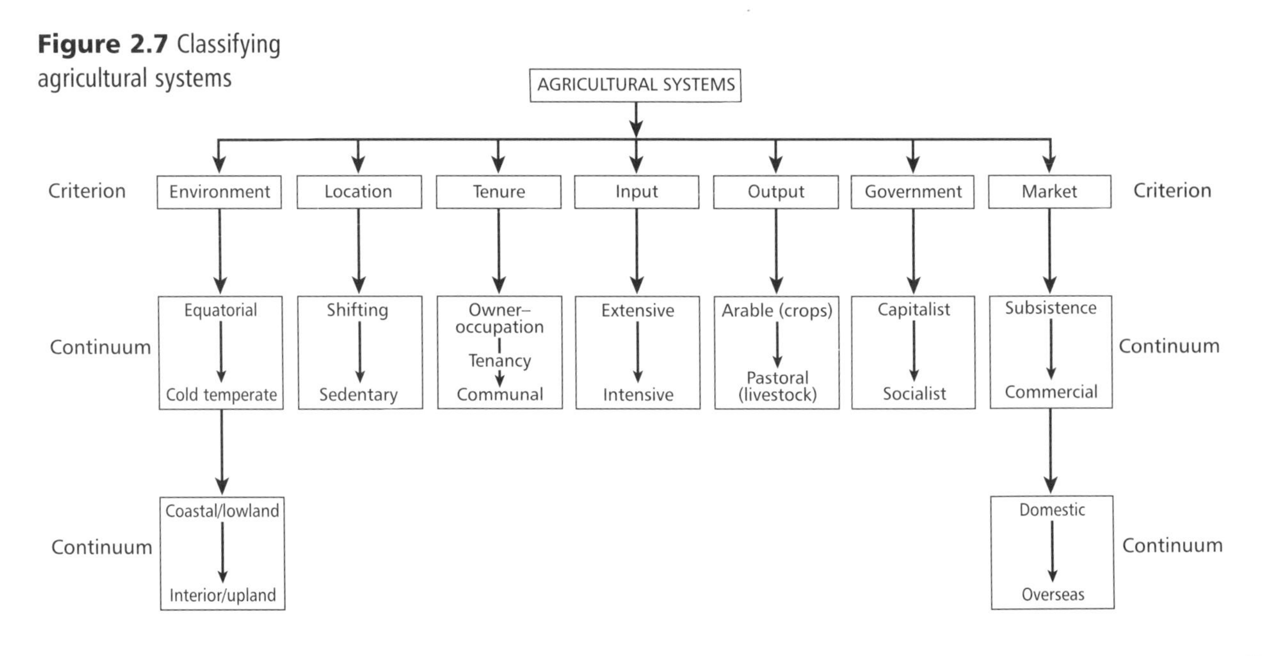

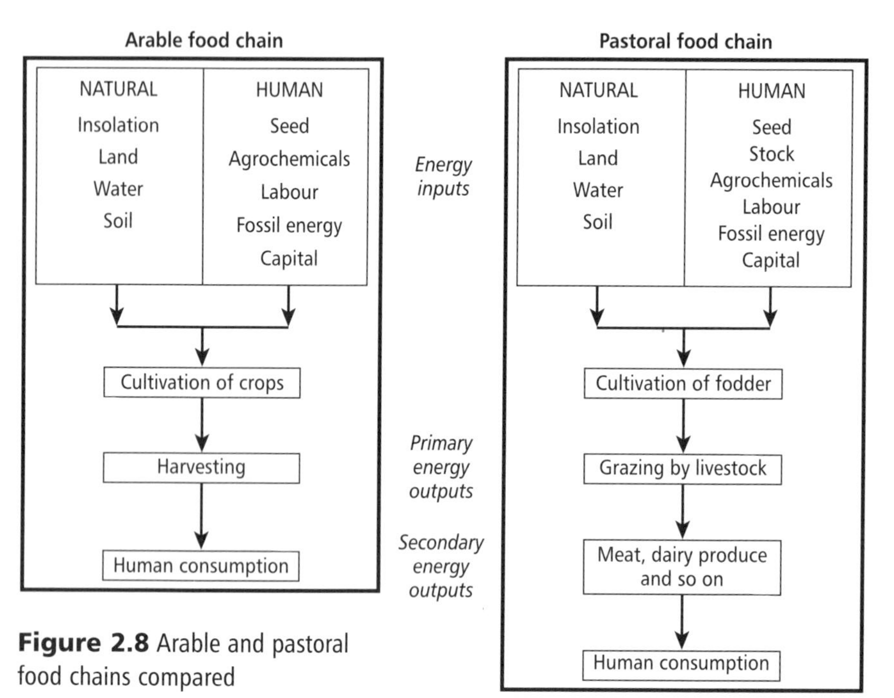

Figure 2.8 Arable and pastoral food chains compared

If the focus is on agro-ecosystems, then environment might be given pride of place. Since climate is one of the major inputs of agriculture, perhaps the distinctions between **equatorial** and **cold temperate** are important. The difference between them, particularly in terms of crops, may well be underlined by other factors, such as the prevailing levels of development, available technology, productivity and so on. Equally, the distinctions between **shifting** and **sedentary**, **extensive** and **intensive**, and **subsistence** and **commercial** may be informative about human pressure at the input end. In each of the three pairings, the latter 'pole' (sedentary, intensive and commercial) points to a higher level of human intervention and therefore also to the existence of agro-ecosystems that are further removed from their original ecosystems.

The efficiency of the agro-ecosystem is another important aspect. By this is meant the relationship of the output to the inputs. How does the output of food (in volume or value terms) compare with the volume or value of inputs? Here the distinction between **arable** and **pastoral** agriculture becomes significant. It could be argued that pastoral farming is a less efficient mode of food production, in that it takes 3 kg of grain to produce a live weight gain of 1 kg in pigs, and 8 kg of grain in the case of cows. True though it is, this is an argument that is rarely voiced by vegetarians (**2.8**).

It is this classification of agricultural types that provides the link to the next chapter. All seven of the criteria used in **2.7** are just as valid and effective when agriculture is looked at in economic rather than ecosystem terms.

Review

14 Which of the criteria in **2.7** do you think is the most important? Give your reasons.

15 Give an example of each of the 18 'polar' agricultural types indicated in **2.7**.

16 How would you classify each of the following:

- rice farming in Japan?
- ostrich farming in Australia?
- organic farming in England?
- fish farming in Scotland?

Enquiry

1 Choose a rural parish close to home or school, and investigate the following topics in the field:
   a What is the dominant type of agriculture? Use **2.7** to classify it.
   b How has agriculture helped to shape today's landscape?
   c Which of the environmental impacts of farming do you think are most serious, and why?

2 Investigate and compare the environmental impacts of one of the following pairs:
   a a large irrigation project in an LEDC and horticulture in a European Union (EU) country other than the UK
   b hill farming in the UK and family subsistence farming in an LEDC.

   It might be helpful to work under the subheadings given in **Section C**.

CHAPTER 3

Farming and the food production chain

SECTION A

An economic view of agriculture

Seeing agriculture as a biological system (an agro-ecosystem) is only one possible viewpoint. The fact that most agriculture is an economic activity means that farms and the whole business of farming may be seen as economic systems. Here the emphasis is more on inputs and outputs of an economic kind, as well as on the economic and political context in which farmers operate. It might help if we agree to use the term **farm system** when adopting this more economic view of agriculture. In this chapter, the spotlight is very definitely on commercial rather than subsistence agriculture. We start by looking at the inputs to the farm system; these may be grouped under four headings.

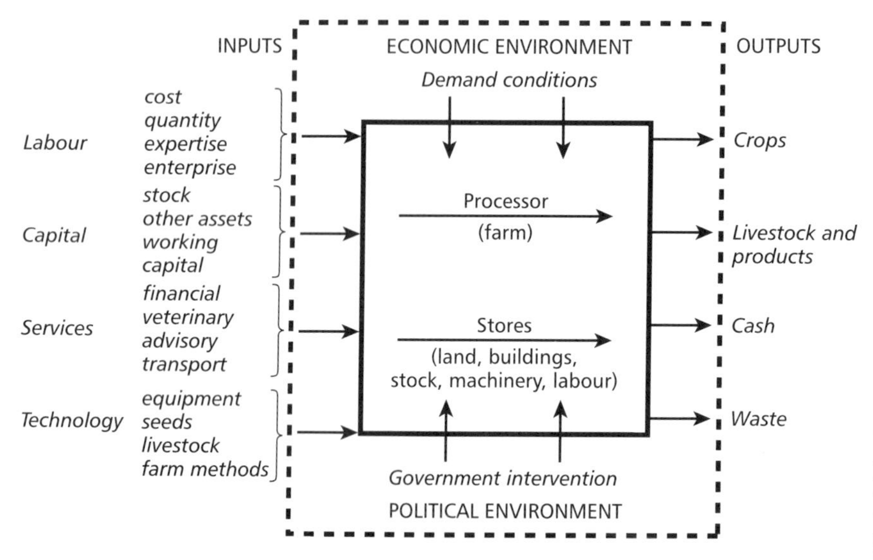

Figure 3.1 The farm system

Labour

There is a vast difference in the labour inputs of LEDC and MEDC farm systems. In most LEDCs, agriculture accounts for well over half of all jobs. Much of the labour requirement is met by the resources of the extended family; much of that labour is female. The information in **3.2** suggests that the decline in the labour needs of agriculture has been minimal. The picture is almost the same in MEDCs, except that the decline as such has been from a low base in 1955. A noticeable feature here is that farm jobs are increasingly part-time. Such is the current state of farming in the EU that nearly one-third of farmers are forced to take on some form of off-farm work. It is in what have been labelled the 'transitional' countries – the newly industrialised countries (NICs), the recently industrialised countries (RICs) and the former socialist states – that the fall in agricultural employment has been most spectacular.

	The percentage of the economically active population engaged in agriculture	
	1955	1995
LEDCs		
Bangladesh	83	62
Cambodia	81	73
Ethiopia	85	86
Madagascar	85	76
Nepal	93	93
Uganda	90	83
Transitional		
Brazil	54	19
Bulgaria	64	11
Romania	70	19
Mexico	54	24
South Korea	55	14
Taiwan	50	19
MEDCs		
Australia	11	5
Canada	11	3
France	20	4
Japan	27	6
UK	5	2
USA	7	3

Figure 3.2 Shifts in agricultural employment (1955–1995)

Case study: A farming exodus

Nearly 24 000 farmers and farm workers in England and Wales left farming in 2000 because of the continuing agricultural crisis. In 1999, the loss was 17 000. During the previous year, the number of full-time farmers fell by 3.6 per cent, that of full-time farm workers by 13.5 per cent, that of part-time farm workers by 12.1 per cent and that of casual workers by 12.0 per cent. This exodus from farming has to be considered against a background of falling prices, declining EU and government support, falling investment confidence and mounting debt (in 2000, farmers owed a total of £10 billion in bank borrowings).

Although the amount of the labour input to farm systems in MEDCs and the transitional economies may be diminishing, mainly due to mechanisation, the costs of it ensure that labour remains a critical factor of production. Likewise, there are still aspects of labour that can never be replaced or supplied by machines. For example, farming involves frequent decision-making. The quality of that decision-making hangs on the experience and expertise of the farmer (see **Section B**). Enterprise is another key quality.

Capital and stock

Capital means all the materials and financial resources required for production. It is not just the cash that is available to buy the inputs of the farm system, such as seeds, livestock and agrochemicals (fertilisers and pesticides). Once purchased, these things become the **stock** of the farm and, together, machinery and farm buildings make up what is termed the **capital assets**. Thus we should distinguish between these assets and **working capital**, which is money in hand for buying stock and paying wages.

Capital is an important discriminator between commercial and subsistence farming. The former is an intensive user of capital, and is sometimes described as being 'asset-rich but income poor'. This is certainly true of farming in the UK today. Low income creates a downward spiral of increased borrowing in order to raise the capital needed to keep the farm system in business. It has been estimated that interest payments on loans now account for something like one-third of British farm income. Perhaps this is where subsistence agriculture scores, for often its capital assets amount to no more than a hoe, an axe and some seeds, while it can function almost without working capital.

Review

1 Referring to **2.1** and **3.1**, highlight the differences between the agro-ecosystem and the farm system views of agriculture.

2 Write a short account analysing the information in **3.2**.

3 Distinguish between a farm's **capital assets** and its **working capital**.

4 Explain why new technology on the farm may be viewed as a 'two-edged sword'.

Services

It is increasingly common for commercial farms to buy in expertise that is necessary to keep the farm system working competitively and efficiently. Financial services, such as accountancy, banking and insurance, are an obvious need, as are veterinary services to help deal with sick livestock. Government departments and companies often provide helpline services to disseminate information and advice on a wide range of matters, from updates on new seed varieties and market conditions to recommendations about fertiliser applications and useful software. Much of the modern machinery found on the farm requires specialist servicing and maintenance. There is often a heavy reliance on transport services, particularly in the collection of farm produce and its delivery to the next stage in the food chain.

Technology

In today's increasingly competitive world, there is much to be gained from exploiting the benefits of the latest technology. It may be the fully computerised grain dryer, new disease-resistant strains of cereal, higher yielding dairy cattle, more efficient mobile irrigation units and so on. But the latest technology usually has its costs. It is invariably expensive, so there is always the need carefully to monitor the costs and benefits and their combined impact on the profit margin. It also has its risks and costs so far as the environment is concerned, as will be discussed in **Chapter 6**.

SECTION B

Decision-making in farming

The point has already been made that the production of food involves the farmer making important decisions almost on a daily basis. In making the 'right' decisions, it is vital that the farmer has a good understanding of the natural environment in which he or she is working – its opportunities and its constraints. The farmer's understanding should also extend to the human or operational environment, to include knowledge of:

- changing market conditions
- alterations to government regulations
- advances in technology.

Then assessments need to be made about the availability of labour, capital and advice. Equally, it needs to be recognised that in a rapidly changing world there are important factors beyond the direct control of the farmer. These may well have an important bearing on the decision-making process.

Figure **3.3** shows how the farmer as a decision-maker has to manipulate a wide range of inputs. These inputs, in turn, affect the character and level of productivity of the farm system. The differences between commercial and subsistence farming become evident towards the bottom of the diagram. The diagram also suggests that, over the years, the farmer acquires

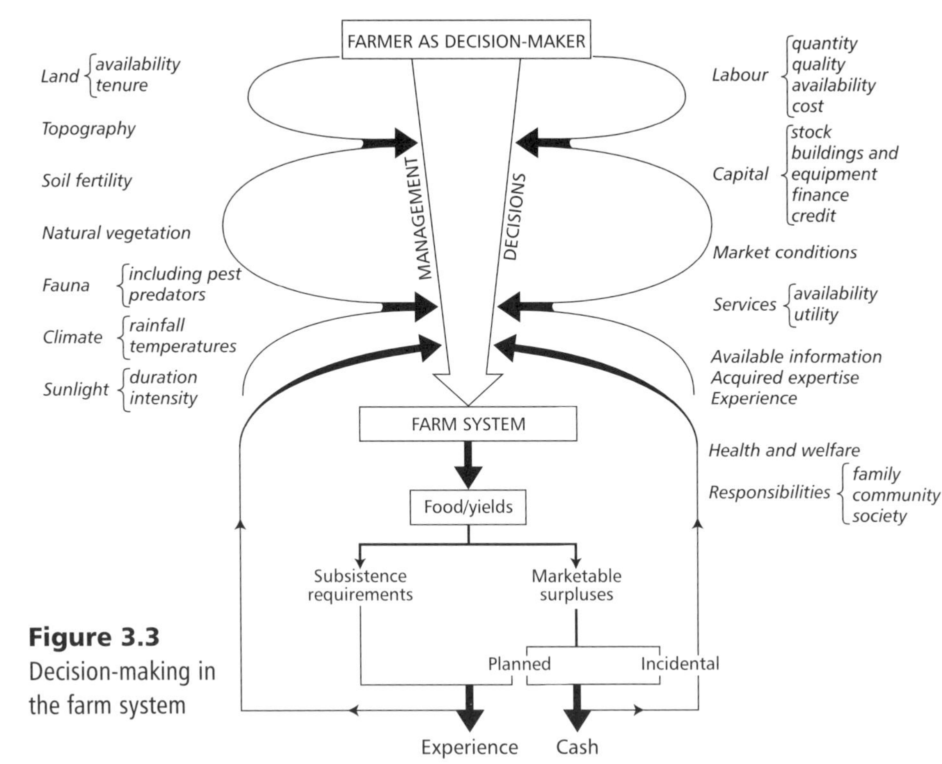

Figure 3.3
Decision-making in the farm system

experience in operating the system. All being well, this experience is turned to advantage and promises sounder decision-making in the future.

Farmers' decisions about how their land is best used are conditioned by physical, economic, social and political factors. These are filtered through the decision-maker's perceptions before the decision is made. Those perceptions are conditioned by at least five different sets of factors:

- personal – age, education, status, income, health
- psychological – values, goals and priorities; attitudes to risk, work and so on
- cultural – traditions, the beliefs and values of the community or society
- information – availability, quantity, quality and the ability to use it
- experience – lessons learnt from past decisions.

The relative weighting of these factors varies with the individual decision-maker and the type of decision being taken.

Optimiser and satisficer concepts

Traditional theories about agriculture and its associated decision-making assumed that the farmer always behaved in a wholly rational way. This **optimiser** concept held that farmers always:

- had perfect knowledge about farming techniques and market conditions
- tried to maximise profits
- sought to minimise costs.

This was the concept underlying perhaps the most famous model in geography – von Thünen's model of the spatial patterns of agricultural production.

Review

5 Which of the factors shown as influencing a farmer's decision-making do you think is most important. Give your reasons.

6 Suggest reasons why and how the personal characteristics listed may be significant factors in a farmer's perception.

7 Explain why information may be problem area in a farmer's decision-making.

8 Summarise the difference between the **optimiser** and **satisficer** concepts.

Although it was, and still is, true that most farmers try to maximise their profits, there are various 'obstacles' or considerations that prevent that aim from being realised:

- **Information** – there is often a problem here because the amount can be overwhelming; it can be inaccurate and the farmer may be incapable of making good use of reliable information.
- **Personal priorities** – there are farmers who do not put maximum profits at the top of their personal priorities. Perhaps some would rank profits below security and low stress levels. No doubt, there are others who are most intent on minimising risk, particularly in situations of volatile markets and uncertain weather.
- **Tradition** – there is the inescapable fact that farm systems are locked in by tradition. There is built-in inertia that inhibits innovation and the switching of a farm's enterprise to meet some passing consumer fashion. For example, if the capital assets of a farm are geared to dairying, this investment would tend to deter a sudden switch to a high-priced crop, even though the potential profits might be much greater. Basically, few farmers can afford to throw away capital assets and many years of experience.
- **Environmental constraints** – finally, and overarching all the above considerations, is the fact that farming is constrained by the physical environment. Freedom to act in pursuit of maximum profits is inevitably restricted by the particular conditions that prevail on the individual farm.

A more realistic appraisal of decision-making is to be found in the **satisficer** concept. This reasons that, although farmers may strive to maximise profits, their lack of knowledge, their inaccurate perception and their personal 'baggage' prevent them from ever reaching this goal. In the circumstances, most are prepared to settle for, and be satisfied with, less.

SECTION C

The role of government

Government intervention is now recognised as a major influence on agriculture and food supply. That intervention can range from a 'gentle touch on the tiller' through steering systems in a certain direction to tight control over both the inputs and outputs of all farm systems. The political spectrum of intervention runs from the capitalist free-market economy to the socialist command economy. But why should governments interfere in what is overtly an economic activity?

Motives

Three main motives are indicated in **3.5**. The strategic motive centres on what is referred to as **food security**. This is to ensure that a nation is able to feed all of its citizens. The only really safe option here is literally to

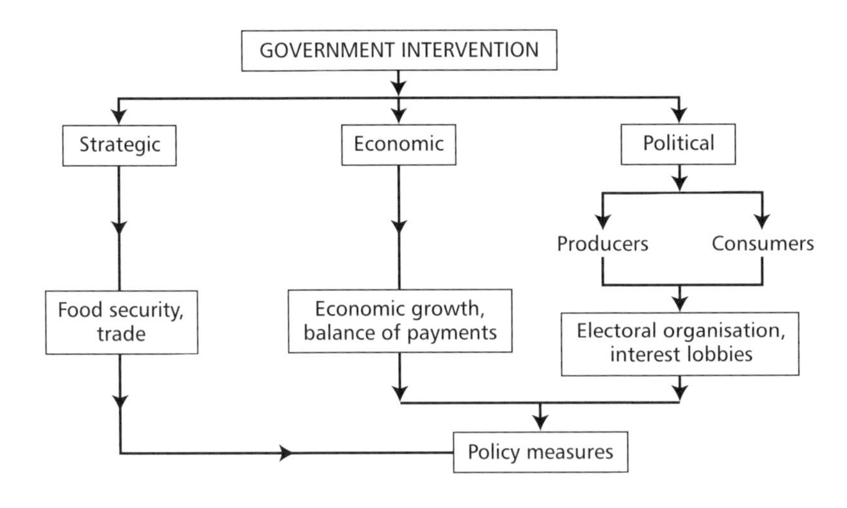

Figure 3.4 Reasons for government intervention in agriculture

produce as much of the nation's food as is possible within its borders. Of course, food supply may be obtained through the medium of trade, but relying on that carries certain risks. It requires that:

- surplus food is available somewhere
- money or exchange goods are available to acquire that food
- the flows of trade will not be disrupted by wars or disputes.

It is fear of the last that has persuaded many governments to strive for high levels of self-sufficiency in food (see **Chapter 4 Section D**). During the Second World War, both Japan and the UK were nearly forced to surrender because blockades on food supplies threatened mass starvation. In general, it is the LEDCs that today worry more about food security than MEDCs. The whole issue is discussed in **Chapters 6** and **7**.

The economic motives for intervention are not far removed from those just considered. However, a number of different viewpoints are possible. One of them might argue that a heavy reliance on imported food can easily lead to an unfavourable balance of trade and a loss of foreign exchange. It is better to supply from home farms than buy in from abroad. Another reason might be that if it is lucrative to produce and export agricultural commodities, then this should be encouraged, perhaps even at the expense of domestic food supply. This is a dilemma that faces many LEDCs (see **Chapter 7**). Yet another viewpoint might argue that governments need to look after agriculture, because it lacks the 'muscle' that manufacturing and services have in the context of economic growth.

The political motives are more to do with party politics and vote-catching. It is a fact that even in some of the most industrialised MEDCs, such as France and Japan, the rural lobby is a strong one. It tends to derive its strength from the number of parliamentary seats supported by the rural vote. Blackmail may be too strong a word, but clearly in such situations the farming community is able to bring much pressure to bear on governments. For certain, there are plenty of instances in which intervention has been triggered by the need to offer 'sweeteners' to farmers. Another often powerful lobby is made up of the food consumers – after all, every voter is a food consumer! There are few governments that are not frightened by unrest over the food supply and food prices.

Figure 3.5 Mechanisms of government intervention in agriculture

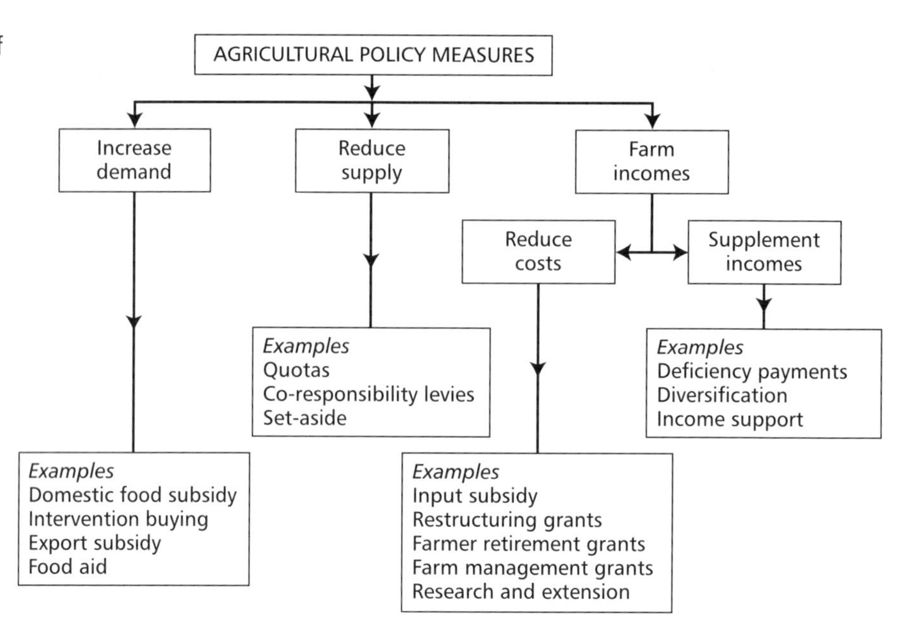

Review

9 What do you think the motives are behind the UK government's intervention in agriculture?

10 Why do you think that there is more government intervention in farming in MEDCs as compared with LEDCs?

11 Argue the case for and against government intervention in agriculture.

Measures

Broadly speaking, a government's agricultural policy measures may have three main aims (**3.5**):

- to increase demand and consequently production
- to reduce supply
- to support farm income, perhaps as part of an attempt to halt the drift of population away from rural areas.

All three main aims are well illustrated by actions taken by the EU and are examined in **Chapters 4** and **5**. For the moment, however, three points need to be stressed at the conclusion of this section:

- Agriculture and farmers, particularly in MEDCs, are now influenced almost as much by the political climate created by government actions as by the economics of the market-place.
- The issue of government intervention is a highly sensitive one, particularly in international discussions aimed at promoting free trade in food. Free trade requires a level playing-field; bias caused by government intervention makes for a slope (see **Chapter 6**).
- All forms of government intervention have a price tag. At the end of the day, it is the taxpayer and the consumer who have to foot the bill. It is something that should weigh on your mind when you start earning!

SECTION D

The arrival of agribusiness

Whilst global agriculture and farm systems around the world are increasingly influenced by individual governments and by groupings of governments – such as the EU and the North American Free Trade Agreement (NAFTA) – there are other developments of no less significance. One such development is the 'industrialisation' of agriculture in many parts of the world, particularly in MEDCs. This is the product of the Third Agricultural Revolution (**1.3**). It represents the latest stage in the evolution of the agricultural curve (**1.6**). At the same time, agriculture is becoming a much more 'capitalised' activity. Money is now being invested in some of its branches on an unprecedented scale. These two processes – industrialisation and capitalisation – have found expression in a newcomer to the agricultural scene, corporate farming, otherwise known as **agribusiness**. In order to better understand this up-and-coming player in global agriculture, we should try to answer three questions:

- What is agribusiness?
- Why has it come into being?
- What are its repercussions?

What is agribusiness?

Agribusiness is a form of agriculture that uses the latest technology (machinery, genetically modified seeds and agrochemicals) to produce cash crops for food, for processing into food and as raw material inputs to other manufacturing industries. It is most commonly associated with large farms and heavy capital investment. In an agribusiness, the organisation of farming is based on scientific and business principles. Often, the farm is responsible not just for the growing and rearing, but also for the processing, storing and packaging of the products, which are usually marketed through major retailing chains such as Marks and Spencer, Tesco and Sainsbury's. Agribusinesses in MEDCs tend to locate:

- in arable lowlands, where large-scale mechanisation is facilitated by huge fields and extensive farms
- within reasonable access of major urban markets.

The 'classic' agribusiness region in the UK is East Anglia.

Why has it come into being?

Traditionally, city investors have shown little interest in agriculture. But that has now changed. Bankers nowadays are keen to lend money to farmers, notably in support of what are described as 'off-land' enterprises, such as horticulture and the intensive rearing of pigs and poultry. They have also been keen to back large farms with their scale economies. Investment has also focused on farm-related activities, such as the 'upstream' production of agrochemicals and the 'downstream' processing of food.

The investment climate for agriculture has changed, with the gradual realisation that food can be a big business as well as a highly profitable one. The basic factor is the boom in food retailing and the ever rising public demand for cheap and exotic foods. The key to success in this highly competitive but lucrative food industry, the key to increased market share, lies in:

- creating food production chains that are controlled by a single company
- capitalising on scale economies
- specialising in particular food products.

It is conditions such as these that have given birth to the agribusiness and to the investors' growing interest in agriculture. Specialisation, in particular, has led to farms becoming less independent and self-contained as economic systems. Agriculture has become drawn into an all encompassing food-producing system that extends well beyond the farm (**1.1**).

There is an interesting contrast between the agribusinesses found on either side of the Atlantic. In the USA, many of the agribusinesses are truly vertically integrated and own most if not all of the food production chain:

- the factories that produce farm inputs such as fertilisers and feedstuff
- the farms that produce the agricultural raw materials
- the factories that process those materials into food
- the distribution networks that market the food.

In Western Europe agribusinesses usually avoid owning farms, but they do own businesses 'upstream' and 'downstream' of the farm, particularly the latter. Instead, they enter into renewable contracts with farmers. Farmers agree to sell to the agribusiness their output of specified products at fixed prices. This is known as **forward contracting**. The price is agreed before cultivation or rearing begins. The agribusiness often specifies the timing of harvest and delivery, as well as the varieties of crop or livestock to be farmed. Such contracts are to the advantage of the agribusiness, as the risks of production are borne by the farmer. Not owning the farm also allows the agribusiness more flexibility to 'shop' around for the best (that is, the lowest) commodity prices. Farmland is a locationally fixed rather than a mobile asset.

What are its repercussions?

The shape of agriculture and the future of rural society in virtually all MEDCs now lie in the hands of agribusiness together with the supermarkets and the food-oriented TNCs (see next section). Widespread forward contracting not only weakens the independence of farmers but it also tends to transfer income from farmers and rural communities to the food-oriented industries. Forward contracting arrangements encourage agriculture in fewer, but larger, holdings; they also encourage still more specialisation. Both allow the agribusiness to exploit more fully the benefits of scale economies. As a consequence, numbers employed in agriculture decline and the farmer becomes less of a manager and more of a caretaker.

Rural landscapes have also been affected by the agribusiness. In Western Europe, field systems have been 'rationalised' and hedgerows and ditches removed, while mechanisation has almost eliminated the need for gang

Figure 3.6 A modern 'industrialised' agricultural landscape

labour (**3.6**). The fields of most agribusiness farms are devoid of human life for most of the year. Factory farming has brought pigs and poultry indoors on a permament basis, while cattle spend many months inside. Indeed, there are now 'zero-grazing' regimes, which may soon mean that fields of cattle will be a thing of the past. Apparently, though, the same industrial production methods do not suit sheep. They still perform better outside and in less sanitised conditions!

More recently, agribusiness has spread to LEDCs, largely through the efforts of the food-oriented TNCs. Here the impact of corporate agriculture threatens to be even more profound, as will be examined in the next section.

Review

12 Try putting together your own definition of **agribusiness**.

13 Why has food retailing become such a thriving business?

14 Explain what is meant by **forward contracting** and why it leads to the leakage of income from rural areas.

SECTION E

Other key players in the food production chain

Thus far, we have identified two influential players in global agriculture today – government and agribusiness. In this section, we look at three more – the TNC, the supermarket and the catering industry. Their influence is most marked towards the end of the food production chain, but such influence also impacts upstream on the farming section of the chain (**1.1**).

Transnational corporations (TNCs)

The distinction between agribusiness and the TNC needs to be clarified. Some TNCs, but only a minority, are involved in agribusiness, whilst there are many agribusinesses that have no links at all with TNCs. For the purposes of this book, the difference is scale. Not only are TNCs bigger businesses, but they operate almost exclusively on an international scale and in economic fields that are not connected with agriculture. Those TNCs with a particular interest in food – production, processing and supply – have become leading players in the **globalisation** of agriculture. The names of such companies may be familiar; they include Unilever, Nestlé, Kraft, General Foods, Coca Cola, Rank Hovis McDougall and Nabisco.

Over the past 25 years, there has been a substantial increase in investment by TNCs in agriculture, mainly in LEDCs. TNCs have been tempted to move in this general direction by:

- cheap land and labour
- appropriate physical conditions (particularly climate)
- improved infrastructure (particularly transport)
- a growing MEDC demand for more exotic foods
- decline in the profitability of investment in other economic sectors.

As a result, some TNCs have increased their involvement in export-oriented agriculture that supplies both fresh and processed food to MEDC markets, as well as the production and distribution of seeds, pesticides and fertilisers.

As a result of this shift in investment, Thailand for example became the major exporter of pineapples in 1979, whereas five years earlier it had not exported the fruit at all. The reason for the change was because the US company Castle and Cooke decided to move a major part of its pineapple operation out of Hawaii. Similarly, the Philippines became a major exporter of bananas by 1975, whereas five years earlier there were no such exports. Again, this was due almost entirely to a shift of investment by a TNC. So, just as those TNCs specialising in manufacturing use **global sourcing**, so too agricultural TNCs have turned to multiple sites of production:

- to lower labour costs
- to obtain year-round supplies of seasonal crops (such as strawberries)
- and to avoid labour and environmental regulations.

Over the last quarter of the 20th century, the global food industry was one of the world's fastest growing industries.

It is in Latin America that this globalisation of agriculture by TNCs has been most extensive and intensive. Of the six countries usually identified as the 'new agricultural countries' (NACs), in which agricultural investment has been concentrated, four – Argentina, Brazil, Chile and Mexico – are in Latin America. The other two are Hungary and Thailand. These NACs are like the NICs in that their governments have encouraged investment in agriculture for the home urban and overseas export markets. That investment is focused on high-value products such as meat, fruit and vegetables. Sometimes the TNCs exercise direct control by purchasing land and supervising production. Increasingly, though, TNCs and international development agencies (such as the World Bank) are encouraging two other arrangements:

- encouraging traditional rural landholders to undertake the mechanised farming of export crops
- contracting out production to peasant producers.

In both cases, the TNC retains control of processing and marketing.

Case study: Nestlé and Unilever – food giants

The Swiss giant Nestlé was founded in 1866 and ranks second only to Unilever in the global league table of TNCs involved in the food and beverage industry. Over 90 per cent of its sales come from food and beverages, with nearly half of its sales in Europe, a quarter in North America and about 10 per cent in Latin America and Asia. It employs about a quarter of a million people worldwide and operates nearly 500 factories in 69 countries.

Food category	Nestlé brand names	Unilever brand names
Beverages	Milo, Nescafé, Nesquik, Vittel and Perrier	Brooke Bond and Lipton
Milk products (including ice cream) and dietetics	Carnation, Chambourcy, Coffee-mate and Lyons Maid	Walt's, Langnese, Ola and Algida, Good Humor and Breyers (USA), Magnum, Solero, Cornetto, Carte d'Or and Viennetta
Margarines, spreads, oils and cooking fats		Bertoli, Puget, Rama and Becel
Chocolate and confectionery	Kit Kat, Polo and Milkybar	
Prepared dishes and cooking aids	Bavarois, Buitoni, Chambourcy, Crosse & Blackwell, Findus, Frisco, Herta, Libbys, Maggi and Stouffer	Birds Eye, Iglo, Gorton's, Ragu, Five brothers, Coleman's, Amora Maille, Upron and Lawry's, Cup-a-Soup, Recipe Soups and Lipton

Figure 3.7 Some brand names in the Nestlé and Unilever portfolios

Unilever, an Anglo-Dutch company incorporated in 1927, is a supplier of consumer goods in the foods, household care and personal product categories. It embraces some 500 companies in more than 80 countries and employs something like 300 000 people worldwide. It is the global leader so far as food products are concerned; this sector accounts for just over half its total business. It also has other operations, the most significant being its plantations: palm oil, tea, coconut and rubber.

Figure **3.7** gives the flavour of the involvement of these two TNCs in the food sector alone. It shows that both have been highly acquisitive in taking over famous brand names. Their fields of operation (both processing and marketing) are truly global.

Finally, we need to take stock of some of the repercussions of this globalisation of agriculture, particularly in LEDCs. Whilst the shopper in a UK supermarket is offered an ever widening choice of foods that are available throughout the year, and at attractively low prices, there is a mounting list of costs in the producer countries:

- TNCs have consolidated control over entire national agricultural sytems.

- Foreign banks have become major agricultural lenders.
- There has been a remarkable expansion of some crops at the expense of others, such as traditional subsistence crops.
- Livestock rearing has led to basic crops being replaced by either pasture or feed-grain production.
- Food staples have been replaced by more profitable products destined for affluent urban and foreign markets.
- Shortfalls in staple food crops have necessitated the increased import of basic food items. This has meant more expensive food and has hit the poorer sections of society.
- There has been an increased concentration of farmland in the hands of capitalist farmers and TNCs. Tenants and sharecroppers have been replaced by agricultural workers, and permanent workers replaced by part-time labourers.
- In order to survive, many remaining peasant farmers and part-time labourers have been forced to borrow money.
- Indebtedness in some areas has persuaded these people to supplement their income by switching to the cultivation of drug crops. The market for such crops in Europe and North America has grown exponentially since the 1970s. The crops can be grown on poor-grade soils and the remoteness of an area can be a positive advantage.

The point has to be made, but not laboured, that in a situation of subsistence farming, there is no need for production chains and companies of the type being described. In the 'hand to mouth' situation of true subsistence farming, there is little reason why the TNC or agribusiness should cast a shadow unless land is being grabbed. However, although subsistence farming prevails in most LEDCs, peasant farmers often indulge in some kind of commercial activity. For example, it might be by:

- producing a small amount of surplus food, which is either sold or traded in a local market
- devoting a small amount of land and time to grow cash crops – these might be fruit and vegetables sold to a nearby tourist hotel, or a crop such as groundnuts that is grown co-operatively and marketed overseas.

Of course, once undertaken, the innocent farmer can quickly be picked up and swept along by the forces of globalisation. That hotel happens to be part of a global chain; the groundnut business happens to be in the hands of a TNC!

The activities of the TNCs also generate uncomfortable ripples for MEDC farmers. Despite the considerable distances between the LEDC producer and the MEDC consumer, the production costs in LEDCs are so low that the MEDC farmer finds it hard to compete on price. Those MEDC farmers who used to enjoy the profits of early season crops, such as strawberries and new potatoes, now find a less profitable market, because global sourcing is able to provide such crops throughout the year and again at low prices.

There is one final aspect of the TNCs that needs to be noted. It is simply the part that they play in moving food around the world. The point was made in **Chapter 1** that, to varying degrees, all countries supplement their domestic supplies with imports of food (**1.1**). Many of those imports are handled, if not provided, by TNCs.

Supermarkets

Wholesalers and retailers are the traditional movers of food (fresh and processed) to the point of sale to consumers. The trend in the second half of the 20th century was for retailers to grow as businesses and for wholesalers gradually to be squeezed out. The processes of change led to the emergence of what are properly termed **multiple retailers**, but which are commonly referred to as **supermarket chains** (Sainsbury's, Tesco, Safeway and so on). Not only have the supermarkets made wholesalers redundant through purchasing most of their food supplies direct from the producers and processors, but they have also caused the number of independent food retailers to fall dramatically. Today, just five multiple retailers handle 65 per cent of the retail food trade in the UK.

Case study: The rise of Tesco

Tesco was founded in 1924. Today, it is Britain's leading food retailer. The founder of Tesco was Sir Jack Cohen. He started selling groceries in London's East End markets in 1919. The brand name of Tesco first appeared on packets of tea in the 1920s. The name was based on the initials of T. E. Stockwell, a partner in the firm of tea suppliers, and the first two letters of Cohen. The first store to be opened was in 1929 in Edgware (north-west London).

Self-service stores started in the USA in the 1930s, during the Depression. It was soon realised that, by selling a wider variety and larger volume of stock and employing fewer staff, such stores could offer lower prices. Self-service stores came to Britain after the Second World War, and the first of Tesco's was opened in St Albans in 1948.

By the early 1960s, Tesco had become a familiar high-street name. As well as groceries, the stores sold fresh food, clothing and household goods. Apart from opening its own new stores, Tesco bought existing chains of stores. This it did during the 1960s and so gradually established a presence over most of England and Wales. In 1967, Tesco opened its first superstore in Westbury (Wiltshire). The superstore was a new concept in retailing – a very large unit on the outskirts of a town, designed to provide ease of access to customers coming by car or public transport (**3.8**).

By 1970, Tesco was a household name. Its reputation had been built on providing basic groceries at very competitive prices. The slogan 'Pile it high and sell it cheap' was the title of Sir Jack Cohen's autobiography. But as people became better off, they started to look for more expensive luxury items as well as everyday household and food products. So, in the late 1970s, the company decided to broaden its customer base and make its stores more attractive to a wider range of customers. Many of the older, high-street stores were closed and the company concentrated on developing bigger out-of-town superstores. While still offering very competitive prices, the emphasis was now on quality, customer service and a customer-friendly environment. In one year in the late 1970s, the Tesco market share increased from 7 to 12 per cent, and in 1979 its annual turnover reached £1 billion for the first time. Although many financial experts did not believe that the company could so radically change its image, the new approach saw sales and profits rise consistently.

Figure 3.8 A Tesco superstore

During the 1980s, Tesco continued to build new superstores, opening its 100th in 1985. In the same year, Tesco introduced its Healthy Eating initiative. Its own brand products carried nutritional advice and many were branded with the Healthy Eating symbol. The company was the first major retailer to emphasise to customers the nutritional value of its own brands.

In the 1990s, the company expanded into Scotland, when it acquired a chain of 57 stores from William Low. By 1995, Tesco had become the largest food retailer in the UK. In the 1990s, Tesco also started to expand its operations outside the UK. In Eastern Europe, it has met growing consumer demand by developing stores in Poland, Hungary, Slovakia and the Czech Republic. Closer to home, in 1997 Tesco purchased 109 stores in Ireland, which gave it a market leadership both north and south of the border.

Retailers such as Tesco need to ensure that customers come through their doors rather than go elsewhere. The leading supermarket chains try to achieve this in a number of different ways:

- by extensive and expensive advertising to promote a brand image, and to persuade consumers that it is to this particular chain that they should come for their food needs rather than any other
- by stocking their own-label products, that sell for prices slightly below those of manufacturers' branded products
- by tempting consumers into their stores with loss leaders
- by trying to undercut the prices offered by their competitors
- by quickly responding to consumer concerns about food additives, GM foods and so on.

Of course, whilst doing all this, the supermarkets have to remain profitable. This they try to achieve by:

- going for high volume turnover
- using computer-controlled, laser checkouts linked to automated ordering systems to ensure continuous stocking of shelves
- setting up carefully organised centralised distribution systems to minimise the use of skilled labour, stocking levels and wastage
- squeezing the prices of fresh produce purchased from farms and of own-label goods produced by outside manufacturers.

The last of these puts the farmer under double pressure as a producer of both fresh food and raw materials for the food processing industry. Both potential buyers apply much of their pressure through the forward contracting system.

Increasingly evident over the last 25 years in the display areas of supermarkets have been the so-called **convenience foods.** The fact that more and more women are going out to work has had a remarkable effect on the food industry. It has created opportunities for the industry to provide less time-consuming ways of preparing meals. Hence the rise of the microwave (cutting cooking time), the cook–chill meal (pre-prepared) and other technologies.

So today multiple retailers have become food marketers rather than distributors of manufacturers' produce. They try to attract customers to their branded premises and then to retain consumer loyalty. But at the heart of these changes is the greater power of retailers to dictate terms to their suppliers, no matter whether they are manufacturers processing food or farmers supplying fresh produce.

The catering industry

It needs to be remembered that the supermarket and street-corner shop are not the only ways in which food finally reaches the MEDC consumer. There are a diversity of outlets serving prepared food, from the hotel to the fish-and-chip shop, from the up-market restaurant to the cosy 'olde worlde' tea-room, from the motorway service area to the truckers' cafe, from the public house to the hot drinks vending machine. All these are part of the so-called catering industry. The fact of the matter is that today more and more food is being consumed outside the home – at school or at work, while travelling or as part of leisure.

Catering falls into two main groups:

- consumer catering, where outlets are open to the public – restaurants, cafes, public houses and so on
- contract catering, where a closed clientele is fed by a contractor or an in-house caterer – as on airlines, in hospitals and increasingly in schools and colleges.

Particularly conspicuous in the first group are the fast-food outlets, both eat-in and take-away. Many of the fast-food chains work on a franchise system. Usually selling one type of meal, such as chicken, pizza or hamburger, they have proved to be highly popular and successful. Factors behind their success include:

- product standardisation
- good market research
- the franchise system, which combines the benefits of local enterprise with those of a large organisation
- the use of cheap, often part-time, labour (a popular if poorly paid student job).

Case study: The Big Mac

McDonald's opened its first restaurant in the UK in 1974. Today, more than 2.5 million people in this country patronise McDonald's every day, either eating in or taking away. Customers trust the company to provide them with food of a safe standard, quick service and value for money. There are currently more than 1000 McDonald's restaurants throughout the UK. They range in location from the high street to the cross-Channel ferry, from the motorway drive-in to the stall at the county show.

McDonald's is the largest and best-known global fast-food provider, with more than 28 000 restaurants in 120 countries. Yet, on any day – and even as the market leader – McDonald's serves less than 1 per cent of the world's population. Behind this market leader, and most other fast-food brands (such as KFC and Burger King) is a vast food processing industry, duty-bound to produce, at knock-down prices, a consistent product that satisfies the food safety standards of the outlet country. Behind the burger-maker is a livestock-rearing network that spans much of the world.

Figure 3.9 Factors energising the food production chain

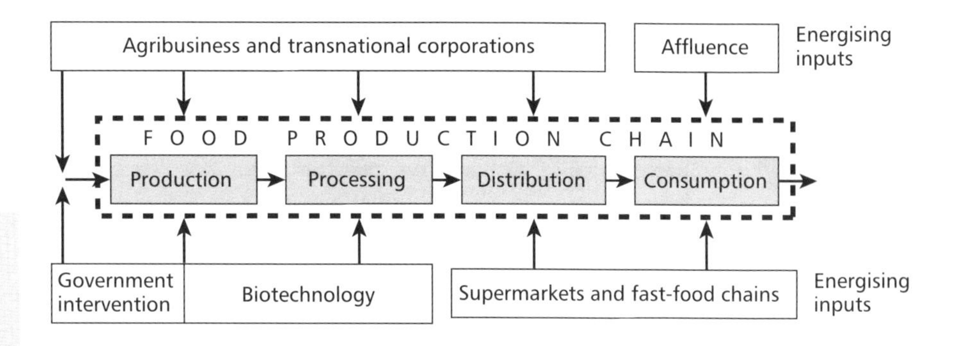

Like the supermarket sector, the catering industry is highly competitive. This again means more pressure being brought to bear on the basic food producers – the farmers.

So, to sum up this rather long chapter, it should be clear that during the last 25 years the food production chains that meet the needs of MEDCs have become more extended spatially (globalised) and possibly more efficient. The chains have been energised at the farming stage by advances in biotechnology and by agribusiness (**3.9**). At the food consumption stage, they have been energised by rising consumer expectations and affluence, by TNCs, by supermarkets and by fast-food chains. The food delivery system in MEDCs has improved immeasurably – it is now faster and offers more choice. However, these improvements have not been without their costs. These have fallen mainly at the farming end of the food chain, both in MEDCs (falling commodity prices) and LEDCs (the TNCs chain increasingly dominating commercial farming). Progress has been mainly for the benefit of the TNCs, agribusiness and the MEDC consumer. There has been little for the LEDCs to cheer about, as should become clear in the remainder of this book.

Review

15 Draw an annotated diagram setting out the repercussions of TNCs on agriculture in LEDCs.

16 Make notes about the benefits of TNCs being involved in the food production chain.

17 Why should multiple retailers wish to cut out the wholesaler?

18 Suggest reasons why food retailing has become such big business.

19 Expand on the reasons given for the success of fast-food chains

20 How many fast-food chains are present in your nearest high street?

Enquiry

Progress in farming is closely linked to the adoption of new farming technology. Investigate the factors that might be expected to affect:

a the spatial diffusion of an agricultural innovation in the North

b a farmer's willingness to embrace that innovation.

Assume that the innovation is either a new cereal strain or an improved breed of livestock. Both promise higher yields and greater disease resistance.

CHAPTER 4

The geography of food surplus and shortage

In this chapter, we return to the theme of the **food–people balance** (FPB) introduced in **Chapter 1**. In development studies, reference is often made to the 'two worlds' of the North (the MEDCs) and the South (the LEDCs). When it comes to food supply and consumption, the same two worlds may be broadly recognised, but with different descriptors – the 'haves' and the 'have nots', the worlds of food surplus and food shortage. The unpalatable truth is that whilst roughly one-quarter of the world's population have far more food than they can eat, the other three-quarters have to suffer varying degrees of hunger. So how glaring are the mismatches between food production and consumption?

SECTION A

Mismatches of food production and consumption

An obvious starting-point in looking for an answer to this question would be a global map of agricultural production. Unfortunately, no such map is available, for the following reasons:

- In many parts of the world, particularly where subsistence farming prevails, few output records are kept – they are hardly relevant in a literally 'food-to-mouth' situation.
- There is no universally agreed way of measuring agricultural output. Should it be by value, volume or what?
- The very diversity of agricultural production makes for difficulties. Should we only take account of food products? Can we measure crops and livestock on the same scale?

Given the absence of a reliable map of global food output, the best we can do is to look at some surrogate measures. There are two good candidates here:

- dependence on agriculture as measured by the percentage of the total population of a country who rely on agriculture for their livelihood (**4.1**)
- the balance of trade in agricultural products (**4.2**).

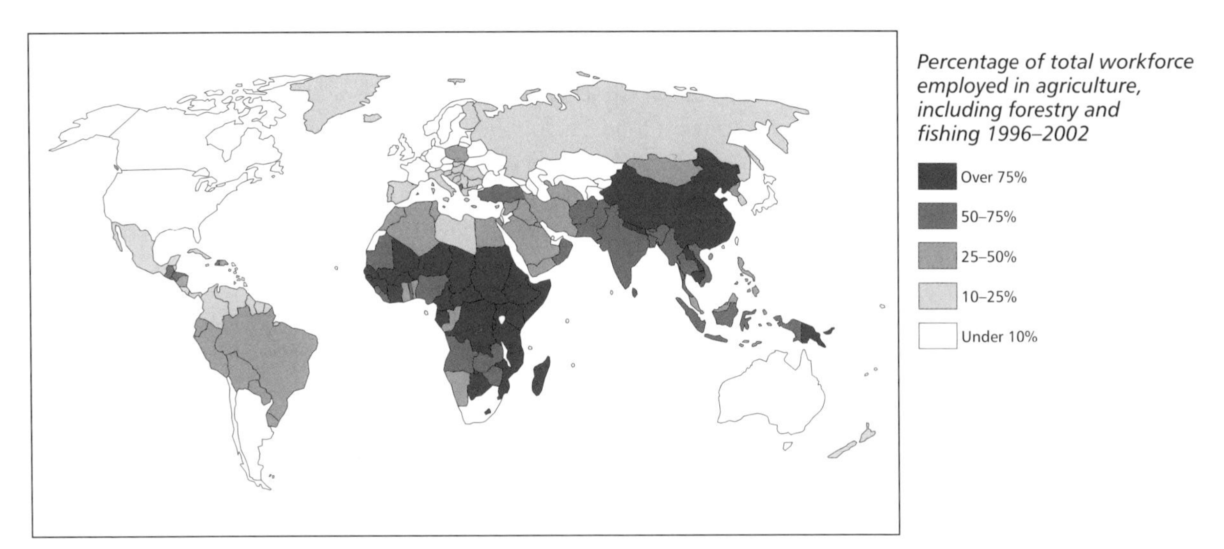

Figure 4.1 Global dependence on agriculture

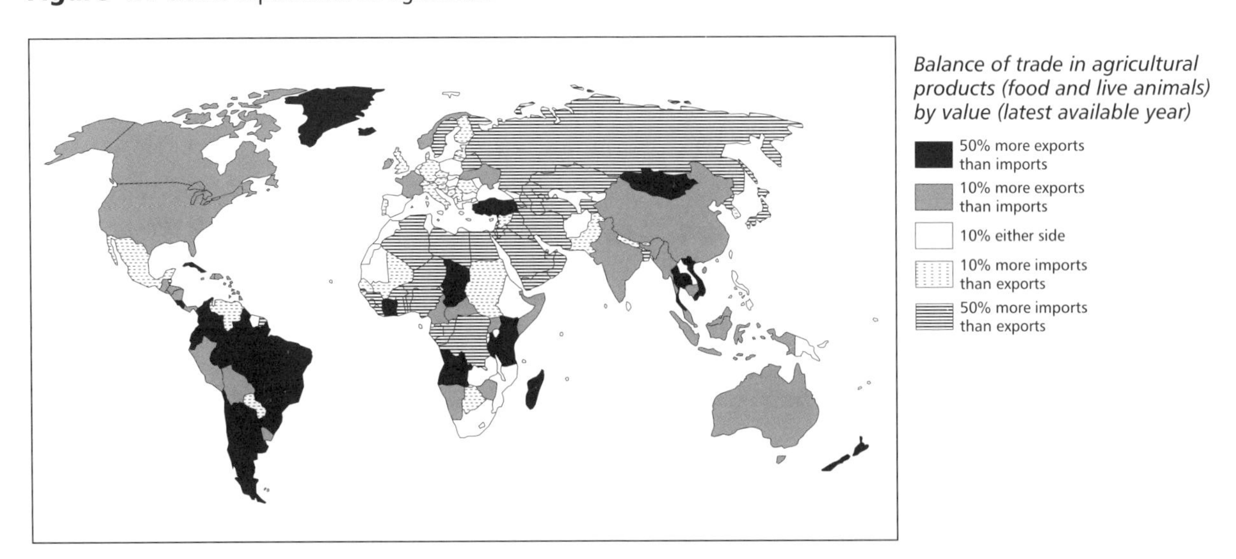

Figure 4.2 Trade in agricultural products

Dependence on agriculture	Trade in agricultural products	
	Exports exceed imports	**Imports exceed exports**
High, > 50%	South and East Asia	Much of Africa
Medium, 25–50%	Central America, parts of South America	The Middle East, Russia
Low, < 25%	North America, parts of Europe, much of South America	Japan, parts of Europe

Figure 4.3 A classification of the status of agriculture

Comparing the two maps allows us to classify much of the world according to a rather crude sixfold scheme (**4.3**). It is the right-hand column that shows a potentially unsatisfactory situation. The picture over much of Africa is one of countries being heavily dependent on agriculture yet unable to produce sufficient food (hence the relatively high level of food imports). Here are most of the countries with the worst food shortages. Where the dependence on agriculture is low, as in parts of Europe and Japan, the reliance on food imports is not a problem. Most of these countries are able to use revenue obtained from other economic activities to pay for imported food. As for the left-hand column, the situation in much of Asia seems to be satisfactory (or is it?). The high dependence on agriculture appears to yield a healthy export trade. It is in the bottom category that parts of the world are identified where perhaps the largest surpluses of food prevail.

Judgements about the adequacy or otherwise of food production need to take into account population, particularly numbers, densities and rates of growth. A look at the world map of population density in your atlas shows that the two regions with the highest population densities are Western Europe and South and East Asia. Neither region was identified as a 'problem' region in **4.3**. A global map of population growth, however, immediately signals possible problem areas, particular where there are high rates of growth. The eye is drawn to Africa, the Middle East and parts of South Asia. It is in the first of these locations that the danger signals flash most vividly. High rates of population growth are occurring in what have already been identified as food shortage countries. At the other extreme, some of the regions with the lowest rates of growth (North America and Western Europe) were previously identified as likely to be food surplus countries.

A map that begins to relate these two dimensions of the FPB is that showing daily calorie intake as a percentage of the minimum requirement for basic health (**4.4**). The frustration here is that it is a good measure in terms of identifying hunger and gluttony, but the data on the map are highly aggregated. A mean value for a whole country is bound to conceal extreme values, and it is those extreme values in which we are most

Figure 4.4 The global distribution of daily calorie intake

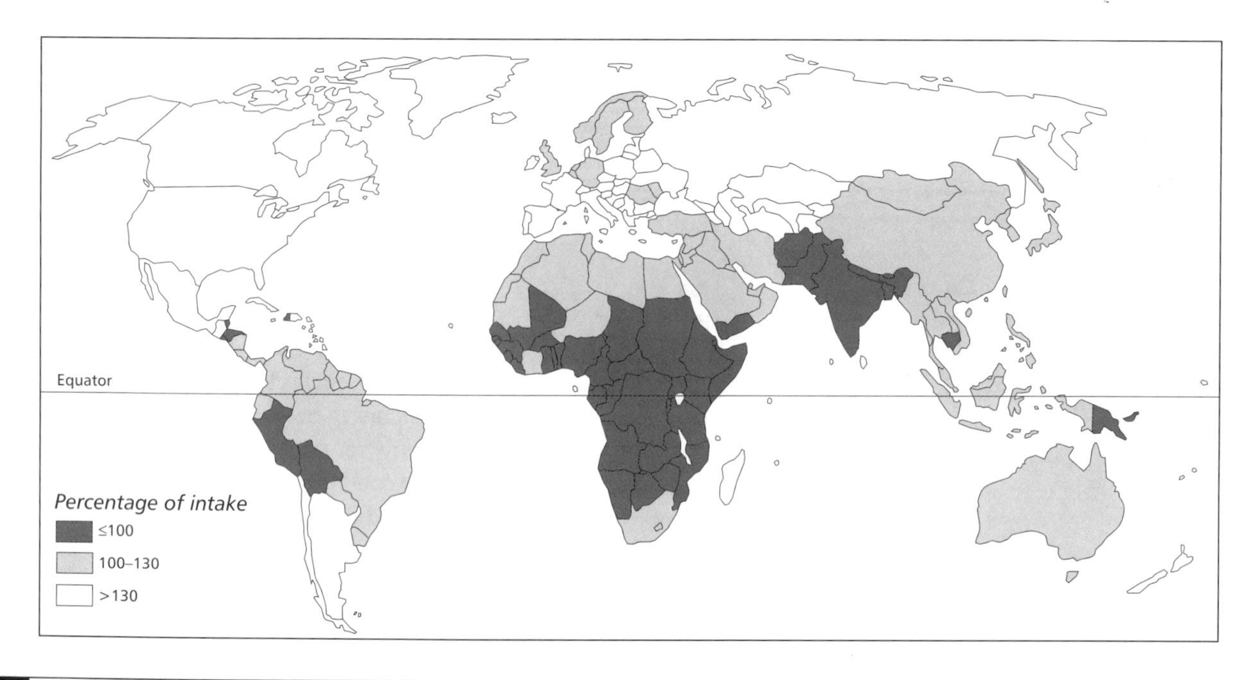

interested. It is estimated that the average daily requirement for people is around 2500 calories. Most people in the North receive more than 30 per cent more than they need, while the average person in the South consumes 10 per cent less. An intake of less than 1500 calories is likely to result in severe malnutrition. It is also estimated that today nearly 600 million people are seriously undernourished.

The map shows that food consumption over much of Africa is below requirement (**4.4**). But also falling in this category is the Indian subcontinent and some adjacent countries. This was not a region that showed up in the preceding maps as one giving rise to concern. Neither did the scattering of six other countries also in this category: Bolivia, Peru, Honduras, Haiti, Cambodia and Papua New Guinea. At the other extreme, North America and much of Europe are revealed as regions where food consumption is well in excess of the minimum requirement. The inclusion of the former Soviet Union in this category is to be queried. During the 1990s, the media frequently carried reports of food shortages and chronic hunger in this part of the world.

It is vital to remember that these global distributions are based on national generalisations. It would be wholly wrong to think that hunger does not exist in the countries of the North. Equally, there are groups within LEDC populations who have more than enough food. Both 'halves' of the world experience inequitable distributions of food.

Review

1 Explain why it is difficult to measure agricultural output.

2 Identify two countries as exemplars for each of the six categories shown in **4.3**.

3 Why should recent rates of population growth be so important in the discussion of food consumption?

4 Select one of the six countries mentioned in the last but one paragraph as falling within the < 100 per cent category (**4.4**). Research the reasons for the unsatisfactory situation.

SECTION B

Related diseases and deaths

It is only in times of prolonged famine that the lack of food is a direct cause of death. However, a persistent lack of proper feeding (say, consuming less than 2500 calories a day) and the associated condition referred to as **malnutrition** can lead to:

- a number of related medical conditions, some of which can prove fatal – these include a range of nutritional deficiency diseases, as well as diarrhoea
- a weakening of the body's resistance to a whole range of communicable diseases unrelated to diet.

It also needs to be stressed that malnutrition can also result from an unbalanced diet. In most cases, this is going to result from some form of mineral or vitamin deficiency, be it A, B or D (**4.5**).

Looking at **4.6**, it is perhaps surprising to find that in Sub-Saharan Africa (the region suspected of suffering the greatest food shortages), the incidence of deaths due to nutritional deficiency diseases is lower than the global average. Although deaths from diarrhoea are up on the global mean, the shortage of food and its sapping of bodily resistance to disease may be blamed for encouraging other killers to head the death charts, such as malaria, respiratory diseases and childhood infections (each responsible for about 10 per cent of all deaths in the region).

Figure 4.5 Some common dietary deficiencies

Iodine
Outcomes – brain damage and mental retardation; an estimated 26 million people are seriously affected; 600 million are affected in some way.

Solutions – add iodine to common salt. Because salt is used in cooking and to preserve foods, sufficient amounts would be absorbed by the general population. Increase the intake of fish, seafood or seaweed.

Vitamin A
Outcomes – an estimated 500 000 children lose their sight each year and half of these children do not survive. Roughly 230 million children have lowered resistance to disease, raising mortality rates by 20–30 per cent.

Solutions – a small intake of green vegetables each day; add the vitamin to sugar or cooking oil; or a tablet, which costs just 1p, to be taken three times a year.

Iron
Outcomes – particularly serious for women in the reproductive age range; impaired work performance; damaged learning ability; dysphagia; pregnancy complications and maternal death.

Solutions – a small adjustment to diet, increasing the intake of red meat, green vegetables or eggs.

Presumably there is a condition of **overnutrition**, which involves not only eating too much but also an unbalanced diet. Eating too much leads to weight gain and obesity. An unbalanced diet – involving, say, too much fat, protein and sugar, or too little fibre – can lead to a whole range of diseases to do with the heart and digestive system. The high values recorded by the 'capitalist economies' of North America, Europe and Australasia under the headings of cancer, cerebro-vascular disease and ischaemic heart disease all have a link to diet and lifestyle (**4.6**). Even the high rate of deaths from neuro-psychiatric (mental) disease may be partly explained by the abuse of alcohol and drugs. The global distribution of deaths from heart disease provides a salutary warning to those of us lucky enough to live in the so-called 'world of plenty'.

Region	Causes of death (% of all deaths)					
	Cancer	Nutritional deficiency diseases	Diarrhoea	Neuro-psychiatric disease	Cerebro-vascular disease	Ischaemic heart disease
Sub-Saharan Africa	1.5	2.8	10.4	3.3	1.5	0.4
India	4.1	6.2	9.6	6.1	2.1	2.8
China	9.2	3.3	2.1	8.0	6.3	2.1
Rest of Asia	4.4	4.6	8.3	7.0	2.1	3.5
Latin America	5.2	4.6	5.7	8.0	2.6	2.7
Middle East	3.4	3.7	10.7	5.6	2.4	1.8
Former socialist economies	14.8	1.4	0.4	11.1	8.9	13.7
Capitalist economies	19.1	1.7	0.3	15.0	5.3	10.0
World	5.8	3.9	7.3	6.8	3.2	3.1

Figure 4.6 Some causes of death, by global region (1990)

Review

5 Distinguish between **malnutrition** and **overnutrition**, and compare their related diseases.

6 Represent the data in **4.6** by means of a suitable diagram. Write a short account highlighting the main messages shown by your diagram.

SECTION C

Scenario 1: Producing and consuming too much (the UK)

Each of the next two sections is devoted to a different food-to-people scenario. Each has as its focus a case-study country. In the first, the UK has been chosen to represent the situation in which the FPB tips in favour of people. In other words, the situation is one of over-production and over-consumption. Both the causes and consequences of this particular imbalance are investigated.

Figure **4.7** identifies the three main causal factors so far as the UK is concerned. What has happened on the people side of the see-saw needs little or no further explanation. But the point should be made that even in a population at or near replacement level, any rise in personal affluence is likely to encourage consumption of foods other than those of the very basic subsistence type (meat, bread and potatoes). The expectation is that the demand for what might be described as the more 'exotic' foods, such as tropical fruits, vegetables and cereals, will increase. Almost inevitably, most such foods will be imported, as will any processed foods with fashionable foreign brand names.

Raising agricultural output

The substantial increase in the productivity of UK farming, particularly on the arable side, has been achieved by a variety of means. These include:

- increased applications of improved fertilisers, pesticides and fungicides
- the development of efficient farm machinery
- enlarging fields to maximise the use of farm machinery
- the 'factory farming' of poultry and pigs
- 'tinkering' with the genes of both crops and livestock to improve their performance and productivity
- the creation of agribusinesses.

Government action

Throughout the 20th century, successive governments have pursued policies aimed at encouraging farming. They have done so for three main reasons:

- To maintain farm incomes at a level that would persuade people to stay on the land and so prevent the decline of rural communities.
- To keep food prices as low as possible.
- To achieve self-sufficiency in basic foods. This was driven by the fear of having food supplies cut off during times of war, as nearly happened during both world wars. It was also driven by the wish to reduce food imports and therefore trade deficits. During the course of the 20th century, the UK's dependence on imported temperate food fell from 67 to 27 per cent.

Government assistance to farming has subsequently been supplemented by help from EU funds. Most help comes in the form of subsidies and grants, and these have been targeted at particular farm products, from sugar beet to lamb. Money has been made available to guarantee the prices that farmers obtain for their products. There have been grants to assist livestock farmers in poor peripheral agricultural regions, such as the Scottish Highlands. Capital grants have helped with expensive agricultural improvements, such as flood defences and land drainage.

Figure 4.7 Factors affecting the food–people balance in the UK

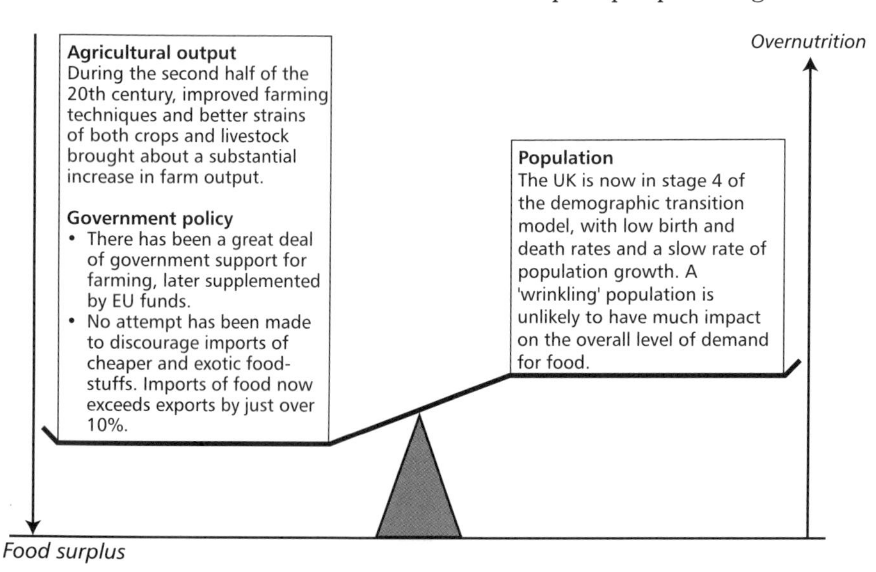

Consequences

The irony of all this help is that it has been too successful, not just in the UK but also in other parts of the EU. The evidence of this lies in 'mountains' and 'lakes' of surplus

1865 (per thousand deaths)		1995 (per thousand deaths)	
Infectious diseases	321	Circulatory diseases	479
Respiratory diseases	148	Neoplasms (cancer)	242
Nervous system	129	Respiratory diseases	108
Diseases of the digestive system	83	Accidents	32
Circulatory diseases	53	Diseases of the digestive system	31

Figure 4.8 Principal causes of death in the UK (1865 and 1995)

production. Despite a long period of intervention pricing, commodity prices have been left to free-fall. They have so collapsed as to threaten the future of much of the EU's farmland. Once again, farmers find themselves in dire straits. As we shall in the next chapter, the official help now being extended to them is of a rather different kind.

Whilst the 'industrialisation' of UK agriculture has resulted in greater efficiency and productivity, it has had major environmental and social costs. The removal of hedgerows to increase field sizes and the reclaiming of marginal land have destroyed wildlife habitats and caused the loss of soil through wind and run-off erosion. Increased use of fertilisers has raised nitrate levels in streams and rivers. Spraying with pesticides and stubble burning close to settlements causes concern in terms of human health. Mechanisation has reduced the need for labour and, by making workers redundant, has encouraged rural depopulation.

For many farmers, cashing in on technological innovations has not always been beneficial. They have been placed on a 'technological treadmill' . In order to maintain their incomes, they have been forced to adopt new methods to increase production and to remain competitive. Not only are the innovations expensive, but so too are their inputs. Often, such costs have run ahead of commodity prices, so that many apparently go-ahead farmers have found themselves heavily in debt.

Finally, it is necessary to return to the theme of the previous section, and to point out another human cost. The evidence in **4.8** suggests that, over the last 130 years, over-consumption has played its part in changing the relative importance of the main causes of death. Perhaps it would be more accurate to say, over-consumption of the wrong types of food. At present, per capita daily calorie intake stands at 3240, which is well above the minimum requirement. But that only places the UK about number 30 in the global rankings; the top spot belongs to Denmark, with a mean value of 3808 calories. In the meantime, the abundance of overweight and over-indulging people ensures that circulatory diseases reign supreme as the number one killer.

Review

7 For what reasons has there been government intervention in the UK's agriculture?

8 Identify the costs and benefits of 'industrialised' agriculture as experienced by the UK.

9 Besides diet, what other factors might help to explain the changes shown in 4.8.

SECTION D

Scenario 2: Producing and consuming too little (Ethiopia)

Here we are looking at the mirror-image scenario of the one just examined (**4.9**). Sadly, there are many countries that might have been selected for case-study purposes. In the event, Ethiopia was chosen, partly because, with per capita GNP at only $100, it is one of the poorest countries in the world and partly because it has been the scene of some recent spectacular famines, such as the one in 1984–1985.

Ethiopia is located in North Africa; it occupies an area of 1.2 million km^2. In physical terms, it consists of two great plateaus separated by part of the African Rift Valley. The Ethiopian Plateau to the west is the most fertile and most densely populated part of the country. East of the Rift Valley is the Somali Plateau, which slopes eastward to the Ogaden Plateau and reaches over 4250 m in the Bale Mountains. The Rift Valley itself, separating the two plateaus, is both deep and narrow, but broadens to the north to form the Danakil Depression, an extensive desert. Arable land accounts for 11 per cent of the land area, permanent pasture 20 per cent and forest 13 per cent.

Ethiopia has a population close to 60 million, so the population density is around 55 persons per km^2; that compares with a figure of 243 persons per km^2 for the UK. In itself, that density figure is not too worrying. Much more so is the fact that Ethiopia's population has just about doubled in the space of 25 years, and this despite at least one major famine. The natural increase rate is currently running at 28 per 1000, whilst the fertility rate is seven children per woman of child-bearing age. Well over three-quarters of the economically active population rely on agriculture for their livelihood.

The question is: Has Ethiopian agriculture been able to sustain this growth in population? With a mean daily per capita consumption currently at 1845 calories, the answer must be 'no' – particularly since this is well below the generally accepted minimum requirement of 2500 calories. The negative answer is reinforced by the fact that mean consumption has been falling. Today, there are only five countries in which the daily intake of food is less: Eritrea (a neighbouring state that was annexed in 1962 by Ethiopia, but regained its independence in 1993), Burundi, Congo, Mozambique and Comoros.

Figure 4.9 Factors affecting the food–people balance in Ethiopia

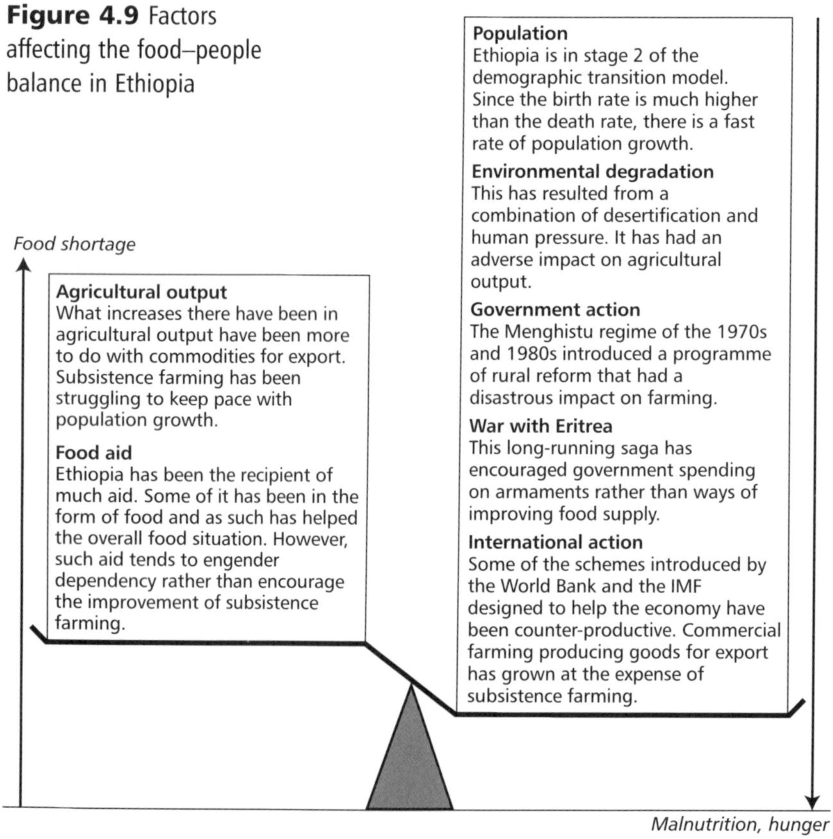

Food shortage

Agricultural output
What increases there have been in agricultural output have been more to do with commodities for export. Subsistence farming has been struggling to keep pace with population growth.

Food aid
Ethiopia has been the recipient of much aid. Some of it has been in the form of food and as such has helped the overall food situation. However, such aid tends to engender dependency rather than encourage the improvement of subsistence farming.

Population
Ethiopia is in stage 2 of the demographic transition model. Since the birth rate is much higher than the death rate, there is a fast rate of population growth.

Environmental degradation
This has resulted from a combination of desertification and human pressure. It has had an adverse impact on agricultural output.

Government action
The Menghistu regime of the 1970s and 1980s introduced a programme of rural reform that had a disastrous impact on farming.

War with Eritrea
This long-running saga has encouraged government spending on armaments rather than ways of improving food supply.

International action
Some of the schemes introduced by the World Bank and the IMF designed to help the economy have been counter-productive. Commercial farming producing goods for export has grown at the expense of subsistence farming.

Malnutrition, hunger

So why has agriculture apparently failed to cope? The answer lies in a mix of reasons:

- The nature of the physical environment bears some responsibility, in that only half of the country is really suited to agriculture. Given the location of the country, with its savanna merging into dry steppe climate, much of Ethiopia now falls within the ever-extending belt of desertification known as the Sahel. As the vegetation is reduced by the increasingly arid conditions, so there is an increasing loss of topsoil by wind and occasionally water.
- It is more than likely, given the high rate of growth, that human population pressure has exceeded carrying capacities, particularly in those areas badly affected by desertification. Once exceeded, a vicious circle of declining agricultural output kicks in.
- Given that crops and live animals make up nearly 70 per cent of Ethiopia's exports, it seems that the more productive land has been taken over by commercial farming (coffee, sugar, yams, fruit and vegetables, as well as cattle, sheep and goats). As a consequence, subsistence farming has been displaced into even more marginal land.
- During the 1970s and 1980s, Ethiopia was a one-party state run by a military government (known as the Mengistu regime). During that time, a programme of rural reform was launched with the aim of spurring agricultural development, increasing food security and addressing environmental problems such as deforestation and soil erosion. The programme included expropriating and redistributing land, setting up peasant production co-operatives, controlling agricultural marketing and pricing, and reorganising rural settlement into a network of villages. The programme proved to be utterly disastrous, worsening rather than reducing the problems. As one observer has remarked, it 'destabilised local food security systems and made the rural population more vulnerable than ever to drought and famine'. The programme was certainly a major contributor to the great famine of 1984–1985.
- Although the notorious Mengistu political regime was overthrown more than ten years ago, the war with Eritrea continues to rumble on intermittently. A large slice of the national budget continues to be spent on armaments and destruction, rather than improving the food supply situation.
- Because of Ethiopia's serious plight, it has been the recipient of much foreign aid. Food aid currently amounts to something like 750 000 tons of cereals a year. However, much aid comes with strings attached. One particularly inappropriate string favoured by the World Bank and the IMF has required the Ethiopian government to increase the prices paid to the growers of export crops, and at the same time to cut support for small farmers and social programmes. The idea was that this would somehow lead to economic recovery. Of course, the opposite has happened. Paying higher prices for export crops has simply led to growing more of those crops at the expense of basic food crops. The external debt of Ethiopia is now equivalent to 160 per cent of GNP.

Review

10 Referring to an atlas, check that you have a clear idea of Ethiopia's location. Draw a sketch map to show the main physical divisions of the country and the main agricultural areas.

11 What are the lessons about food supply to be learnt from the Ethiopian case study?

12 Explain why, in **4.9**, government reform, war with Eritrea and international aid should be shown as adding to the population end of the see-saw?

In representing Ethiopia's FPB in **4.9**, the last three factors just discussed – government reform, war with Eritrea and international aid – are shown at the population rather than the agricultural end of the see-saw. No matter what their intentions may have been, these actions have not helped agricultural output and therefore they are to be seen, along with desertification, as adding to the general burden of population. No one doubts that there are still huge numbers of hungry people in Ethiopia. Government officials argue that Ethiopians 'are hungry because they have no money, not because there is a lack of food'. Do they not realise that hunger and poverty are almost inseparable twins? It seems obscene that poor countries, such as Ethiopia, have millions of hungry people and yet are exporting food to countries where people are already well-fed!

Attempts to rectify the two imbalances scrutinised in the last two sections will be examined in the next chapter.

SECTION E

Seeking food security

Perhaps in an ideal world, every country should strive to become as self-sufficient as possible with respect to food supply. This requires reaching a state of equilibrium where food supply and food demand are nicely balanced. In short, the FPB needs to be horizontal rather than inclined, as in **4.7** and **4.9**. The provision of all or most of a nation's own food needs offers what most countries seek – food security. **Food security** may be defined as the ability of a country to guarantee an adequate supply of food for all its population.

So which countries today appear to be most food secure? They are not necessarily the MEDCs, with their so-called food surpluses. Many of these have a high dependence on imported food. For example, 15 per cent of Japan's imports are made up of food; for the UK, the figure is 8 per cent. Dependence on imported food requires earning sufficient foreign exchange to be able to make the necessary purchases. Dependence on imported food also requires securing reliable supplies. Both of those conditions have risks attached. So perhaps the best food security is that rooted in domestic agriculture. In other words, the most food-secure countries may be those whose overall food supply depends least on imports. Unfortunately, data that analyse national food supply in terms of these two components – domestic and imported – are not widely available. Instead, we are forced to use a surrogate measure: the percentage importance, by value, of food in a country's imports. It is not perfect, but at least figures are available for each and every one of the world's nations.

Figure **4.10** lists the ten countries with the smallest percentage of imports given over to food and livestock. What do we find?

- Three of the countries – Turkey, Australia and Thailand – also seem to be in business as exporters of food.

Country	Food and livestock	
	Percentage of all imports	Percentage of all exports
Turkey	2.0	18.2
China	2.7	8.3
Singapore	3.3	2.4
Australia	3.6	13.2
Thailand	3.7	20.6
South Africa	4.0	5.7
Malaysia	4.3	2.9
Taiwan	4.5	0.9
South Korea	4.6	2.4
Austria	4.8	2.9
UK	8.2	4.6

Figure 4.10 The top ten most food-secure countries (1995)

- Four of the countries – Singapore, Malaysia, Taiwan and South Korea – are Asian 'Tigers', who have achieved their developmental status largely by growth in the secondary (manufacturing) and tertiary (services) sectors.
- The world's most populous nation, China, ranks as possibly the second most secure nation. Can this really be so?

Two of these countries, namely China and Taiwan (formerly part of China, before the creation of the People's Republic in 1949), have been selected for closer scrutiny. Is there anything to be learnt from the 'giant' and its 'dwarf' neighbour that might help other countries keen to improve their level of food security?

Case study: China

China accounts for 20 per cent of the world's population, but only just over 6 per cent of the world's land area. Even so, it has managed to farm this area in such a way as to just about feed its immense population, which now stands at over 1.25 billion. But China is on a knife-edge, in that it only needs a slight shortfall in the expected rice or wheat harvest and the country is plunged into a situation of hunger, malnutrition and even famine. A persistent theme throughout much of China's history has been the uneasy balance between food production and rising population numbers. Since the Communist takeover in 1949, there have been two major reforms of Chinese agriculture:

- The Great Leap Forward, launched in 1958, started the 'collectivisation' of Chinese agriculture. It was assisted by: the use of new higher-yielding seed varieties of rice, wheat and maize; heavier applications of fertilisers; and more irrigation and intercropping (growing two or more different crops together). Production of rice and wheat did increase, but not sufficiently to increase per capita food consumption.
- After 1978, rural and agricultural organisation was changed yet again. The collective fields were broken up and the land distributed on long-term leases to peasant households. Although the state still guided agricultural production through setting production targets, quotas and prices, a 'free-market' system was allowed to run. This has increasingly shaped what peasants grow and how they market their produce. At first, there were significant increases in production, but the rises have not been sustained.

The painful bottom line is that the threat of famine has not receded in China. Indeed, between 1959 and 1961 it suffered one of the worst famines known in human history, which resulted in between 25 and 30 million deaths. Flood and drought both played a part in that famine, but the bulk of the blame fell on the government, because it tried to push the collectivisation programme through too quickly. Another famine threatened in 1981 after the failure of the monsoon rains, but it was averted by accepting international aid and buying grain on the world market. Since then, China has been on the brink of famine several times. Whilst agricultural production continues to rise in a faltering way, the cutback in the rate of population growth through strict birth control policies has helped to redeem the situation a little.

Case study: Taiwan

Although part of China up to just over 50 years ago and similarly geared to rice production, capitalist Taiwan now has a markedly different agricultural system. Between 1948 and 1953, the Taiwanese government undertook a programme of radical land reform. This involved an equitable redistribution of land among family farmers. Economic aid from the USA at the time ensured that these farmers had the capital to invest in fertilisers, machinery, irrigation equipment and improved strains of rice and other crops (**4.11**). Agricultural productivity increased significantly, and the associated profits were creamed off to finance the industrial development of the country. As industrialisation has progressed, so the agricultural sector has become less important. Even so, and despite considerable rural–urban migration, the country is able to produce most of its own food requirements. The principal crops are rice, sugar, fruit and vegetables; beef and pork are also key products. Very little of these commodities enters the export trade (**4.10**).

Figure 4.11 Highly mechanised farming

Current government policy aims to protect Taiwan's agriculture from competitive and fluctuating world market prices. At the same time, it is encouraging a second phase of land reform in which small family farms are being converted into large, mechanised commercial units. All in all, it looks as if Taiwan now enjoys a good measure of food security. This achievement becomes all the more impressive when it is realised that its population has almost trebled since 1950.

Three important messages emerge from these two case studies:

- No matter whether they are communist or capitalist, it should be the responsibility of all governments to ensure that they accurately know what is required adequately to feed their people. They also need clearly to determine what is the best way to secure that supply: what should be raised domestically and what might be imported.
- Except in those cases in which the physical inputs (such as heat and water) are unreliable, it makes strategic good sense to rely as much as possible on domestic rather than imported food supplies. Dependence on imported food can easily be threatened by uncertainties: international politics and global commodity prices to name but two.
- True food security is more than just producing the minimum needed to feed the national population. Ideally, it requires some surplus production to cushion against crop failures and other disasters. When it comes to agriculture, no country is truly risk-free.

This chapter has examined three different scenarios with respect to the food–people balance. The two extreme situations of food surplus and shortage are likely to be repugnant to most observers: the one because of its implied greed and waste, and the other because of its deprivation and injustice. The hope is that food surpluses can be moved to countries in need (**Chapter 5**) and that most nations can find some sort of food-secure situation that is midway between surplus and shortage. But the closer one moves to this midway point, the more finely balanced situations are. It only needs some small event outside the control of the food producers and the whole FPB can easily and quickly tip towards disaster.

Review

13 Explain why food security is so important.

14 Why is it better for a country to rely on domestic rather than imported food?

15 Do you think that the physical extent of a country is a problem or an advantage when it comes to food supply? Justify your answer.

Enquiry

Investigate and compare the food security situations in one of the following pairs of countries: Japan and Uruguay; Argentina and Zimbabwe; or the Gambia and Thailand. Pay particular attention to:

a the relative importance of food in both exports and imports
b rates of population growth
c government actions.

CHAPTER

5 Responding to food surplus and shortage

In this chapter, attention turns to efforts that have been made in order to correct some of the disparities in the global distribution of food production and so bring about a better balance between food and mouths.

SECTION A

Producing less in the North

The mountains and lakes shown in the cartoon (**5.1**) represent the enormous success in terms of food security of the EU's Common Agricultural Policy (CAP). By 1973, the European Common Market (as it was then known) was nearly self-sufficient in terms of cereals, beef, dairy products, poultry, vegetables and much more. The CAP, introduced in 1958, was specifically designed to:

- increase the productivity of agriculture (it was in a bad state)
- promote self-sufficiency in crops such as sugar and oil seed (much needed to be done here)
- stabilise markets for farm products (demand and prices were volatile)
- provide a fair standard of living for EU farmers and their families (many, especially in Southern Europe, were below the breadline)
- ensure that food was available to EU consumers at a fair price (after the Second World War, prices were often high due to shortages).

Phase 1 of the CAP was centred on a system of guaranteed prices to support farmers. In addition, EU farmers were subsidised to make their products competitive on the world market. Imports into the EU were subject to import duties or levies to protect its farmers. A similar scheme of farm support existed in the USA. Inevitably, LEDCs found that this created a sloping 'playing-field'. They were forced to take cheap imports from the EU. At the same time, they found it very hard to export to the EU, because of the barriers created by the import duties.

Successful as the CAP was in solving the issue of EU food security, it was already clear by 1973 that it had to be reformed. Two-thirds of the EU's budget was being spent on it. The relentless treadmill of intensification led to huge surpluses that were expensive to store and difficult to dispose of. The Common Market was also rocked by a number of CAP fraud scandals.

Phase 2 of the CAP, from 1979 to 1991, represented the first attempt to cut back on production. Milk quotas were introduced to control the size of the milk lake and butter mountain. In 1988 Set Aside was established, initially

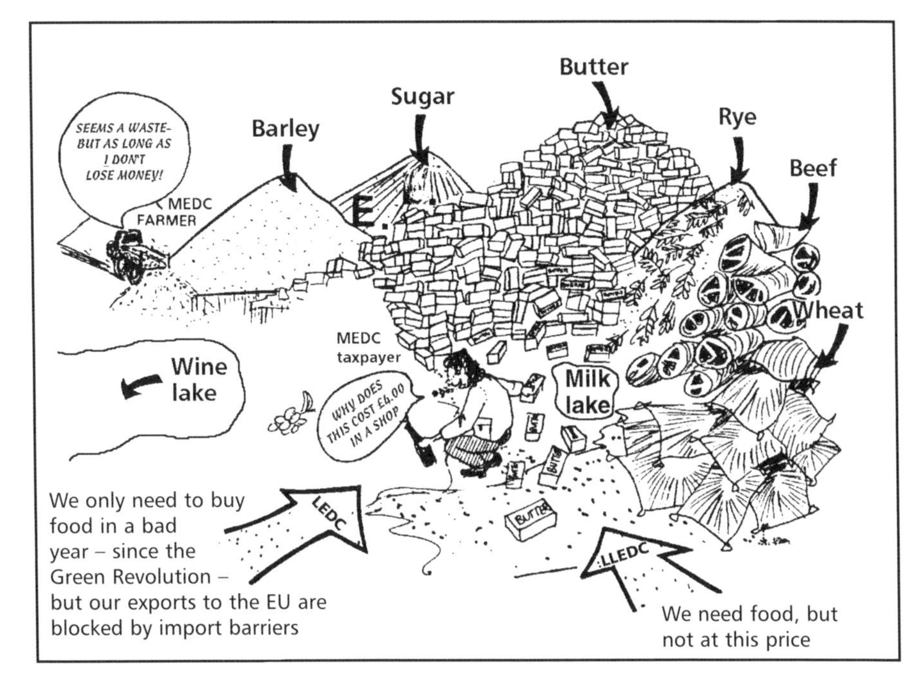

Figure 5.1 A cartoon about EU surpluses

on a voluntary basis. It allowed farmers to be paid for taking up to 20 per cent of their land out of production, to be left fallow, converted to woodland or used for non-agricultural purposes.

Phase 3 covers the period from 1992 to 2000, when there was a growing awareness of the huge environmental costs of agricultural intensification. There was also increasing public resentment about the costs of subsidies to farmers (it was costing each EU citizen £250 a year). So the key elements of this phase were price cuts and withdrawal of land from production, but with more emphasis on agri-environmental schemes and farm diversification.

Phase 4 of the CAP, known as Agenda 2000, developed for a number of reasons. These included:

- The projected cost of the CAP with the enlargement of the EU to include the substantially peasant economies of Eastern and Southern Europe.
- The refusal of Germany to continue to provide so much money for the CAP. This would hasten the need for further cost cutting by less direct support for farm prices.
- The pressures of the World Trade Organisation (WTO) for countries to stop subsidising their farming industries. In 1998, the USA had done precisely this with its Freedom to Farm strategy, whilst New Zealand had successfully shown the way to develop a subsidy-free farming industry.
- The increasing health concerns, post-BSE, about the quality of farm produce and the way in which it is produced (the animal welfare lobby). This was typified by a growing interest in 'natural' farming and chemical-free food.
- The renewed emphasis on conservation of the countryside. The destructive nature of agro-chemicals on wildlife and of herbicides on hay meadows and hedgerows, the spreading menace of soil erosion, and fast overland flow leading to flooding have all been much publicised.
- The recent farm crisis in the UK, made worse by the high value of the pound Sterling against the Euro and the outbreak of 'foot and mouth' disease.

The **farm crisis** in the UK and other parts of the EU, brought about by surplus food production and plummeting market prices, is threatening to turn into a **rural crisis**. This is especially the case in the hill-farming areas, which have become so reliant on support mechanisms. The powerful countryside lobby has forced the UK government to develop overall rural strategies that integrate farming into a wider rural context. Therefore, whilst Agenda 2000 recommended price cuts of 10 per cent for milk, 20 per cent for cereals and 30 per cent for beef, it also boosted a whole range of agri-environment support grants and farm diversification strategies, such as the England Rural Development Programme.

The case for continuing government support of the farming community in the UK is based on the fact that:

- although only 1.7 per cent of the population is employed as farmers, they look after most of the land – they are the custodians of the countryside
- farmers purchase a range of seeds, tractors, fertilisers and basic goods, so a crisis in farming will hit a lot of supporting businesses (**1.1**), thus creating a de-multiplier effect.

It is argued by many rural people that a national government whose traditional core support comes from urban industrial areas understands neither farms nor the wider rural issues.

At this moment in time, it seems that there are two courses of action open to the EU and its member governments to stave off the farm and rural crises:

- Work existing farmland in a different way. There are two obvious possibilities here: **extensification** (adopting organic farming and other environmentally friendly modes of farming); or growing more non-food crops (for example, biomass for energy generation).
- Take land out of agricultural production: restore wildlife habitats; create new woodlands and forests; or use it for recreation.

Review

1. Produce an annotated diagram summarising the evolution of the CAP.
2. Assess or debate the arguments set out in **5.2**.
3. Find out more about the England Rural Development Programme, 2000

Figure 5.2 Some arguments for and against CAP subsidies

For	Against
■ There are no practical alternatives to subsidies. ■ Without subsidies, many upland and marginal areas would 'go under' as a result of mass out-migration and economic collapse. ■ Support has recently led to farmers developing more environmentally sustainable modes of farming. ■ Farmers are willing to change, but they need protection from the associated risks.	■ Subsidies inflate the cost of land and other inputs, thus forcing farmers on to the technological treadmill. ■ Subsidies encourage land degradation. ■ It is the large, intensive farms that have benefited most from support. ■ The countryside would be kept in better shape were there no subsidies. ■ It would be much fairer to the LEDCs if MEDC farming were not subsidised.

Go organic and diversify

Organic farming is the UK's fastest growing branch of farming. In 1993, the organic market was barely measurable, but it has since been growing exponentially from this small base by up to 40 per cent a year. It can barely keep up with demand. A recent survey in the UK indicated that 80 per cent of shoppers had bought organic at some time or another. Organic food is now a global industry. Austria, the Netherlands, Italy, the Scandinavian countries, the USA and Australia are all pouring money into research, and offering grants to support conversion to this mode of farming. In Germany, which is just experiencing its first BSE scare, the government is saying that industrial agriculture must end, and the Agriculture Minister (a member of the Green Party) has set a target that 20 per cent of total output should be organic by 2010. In Denmark, 20 per cent of all produce is already organic, and there is a thriving export trade in such produce. Not so long ago, organic farming was seen as the province of a few 'health-food freaks', but now it is very fashionable – just like wanting better health and education.

Why is this? The following are six reasons frequently quoted in support of organic farming:

- it produces healthy food, full of natural flavour
- it does not involve the use of GM materials
- it require farmers to care for the countryside and water resources
- it is humane to animals
- it is good for the environment, using mineral pesticides and less inorganic fertilisers
- it supports local enterprises and rural economies.

Perhaps most powerful of all is the overarching view that organic farming is sustainable, in that it seeks to ensure the future of our food, our health and the environment that we hand on to the next generation.

Case study: A blueprint for the future of the hills

Vyrnwy Farm consists of nearly 5000 ha (that is huge by UK standards) of marginal rough grazing in the hills of mid-Wales. The farm lies in the catchment for Lake Vyrnwy, the major source of drinking water for Liverpool. It is thus located in a highly sensitive environment, with rigid public health and anti-pollution rules to ensure that farming does not contaminate water supplies.

The land is owned by Severn–Trent Water, who have invested in leading-edge technology for the production of cattle and lambs. They work in partnership with the RSPB, who manage the farm on a day-to-day basis and share both the profits and losses. The RSPB aim is to manage the farm in an environmentally friendly way so that birdlife may recover from the decimation caused by sheep overgrazing. The RSPB manager is a local farmer.

Severn–Trent invested in cattle housing with an automatic feeding system, and straw- bedded yards to cut pollution run-off. They also invested in sheep housing with an open-sided design to hold 1200 ewes, and a revolutionary sheep-dipping system capable of holding 1500 sheep, which prevents pesticide pollution. The housing enables the farmer to take the sheep off the fells in winter, thus allowing the pastures time to recover. The state-of-the-art dip avoids farm pollution, as do plastic waste containers. To add value, the lambs are clipped before sale.

Fodder is imported from a number of lowland farms, all of which use organic methods. The target is for the fully organic production of beef and lamb by August 2001. Farming returns are up, the environment has improved and birdlife has increased. It is a real success story of leading-edge technology working with the environment.

This case study clearly reinforces some of the positives of organic farming. So far it seems to be all good news. However, the next case study shows that there are considerable costs in going organic. That is why it is vital for farmers to have start-up grants to help them convert to organic farming.

Case study: A farmer counts the cost of going organic

Geoff Goodman, his wife, their son and one full-time employee work a mixed farm of 880 ha. Recently, Geoff called in a consultant to advise on the feasibility of converting to organic methods. At present, there is a dairy herd of 145 cows, 63 beef cattle and a free-range flock of 1800 geese; 50 ha is used for growing wheat, barley and fodder crops.

The consultant's figures showed that whilst money would be saved on fertilisers and chemicals as a result of relying on crop rotation to maintain soil fertility, Geoff would need to spend £40 000 on reseeding pasture, on extra storage for grain and manure, and on new machinery for spraying muck and weeding unsprayed fields. More worrying was the estimate that full organic status might take five years to achieve. During that time, farm output would be falling because organic yields are lower, and he would be not yet be enjoying the compensation of higher prices paid for organic products.

A further concern is that as the organic bandwagon 'rolls' and 'organic' becomes a global brand, the local producer-to-consumer linkage will be broken, especially if the big supermarkets move in. In Britain, 60 per cent of organic food sales are to less than 10 per cent of the population (largely social class A/B). This is mainly because of the higher production costs and therefore the higher prices of such food. If the supermarkets take over organic food sales, then such foods may be expected to hit a wider market. But there is a worry that as they take over the organic food business, the supermarkets will 'dilute' the organic brand image. This will be by claiming organic ingredients in convenience foods and by transporting organic produce halfway round the world at massive, but hidden, environmental expense. The organic global food basket is unlikely to benefit the LEDCs.

Organic farming is just one example of extensification; that is, farming less intensively. Other agri-environment schemes also do this, as for example the designation of environmentally sensitive areas where there are strict controls on the use of fertilisers and water resources, as well as on the disposal of farm wastes and slurry.

The second way of tackling the farm crisis is to take farmland out of agricultural production altogether and to use it for something else. The need for farmers to do this has escalated as farm incomes have crashed due to the falling prices of cereals, vegetables, beef, sheep and pigs. Currently, all is doom and gloom for British farming. In 2000 alone, 22 000 farm workers lost their jobs and many farms, especially hill farms, traded at record losses. The BSE crisis and the outbreak of 'foot and mouth' disease have added to the tale of woe. **Diversification** is one way of getting more profit from the land and it has become essential for most farmers. Indeed, at present the 'sideline' is the now the main income for many farmers.

As can be seen from **5.3**, farm diversification consists of two basic strands:

- agricultural diversification – broadening the range of farm output
- structural diversification – using the farm and land as a base for non-farming activities.

The precise nature and likely success of any diversification depends on:

- the conditions laid down in planning permissions – this is a major problem in green belts, national parks and AONBs
- the availability of start-up funds, capital and expertise to plan and market the business – grants can be very influential
- the location of the farm – some activities require passing trade and proximity to urban areas, whilst others require remoteness
- the current state of the market (demand, competition and costs) – this can easily hit the viability of alternative activities
- the compatibility of the new activity with established farming practices
- the innovativeness and co-operation of farmers.

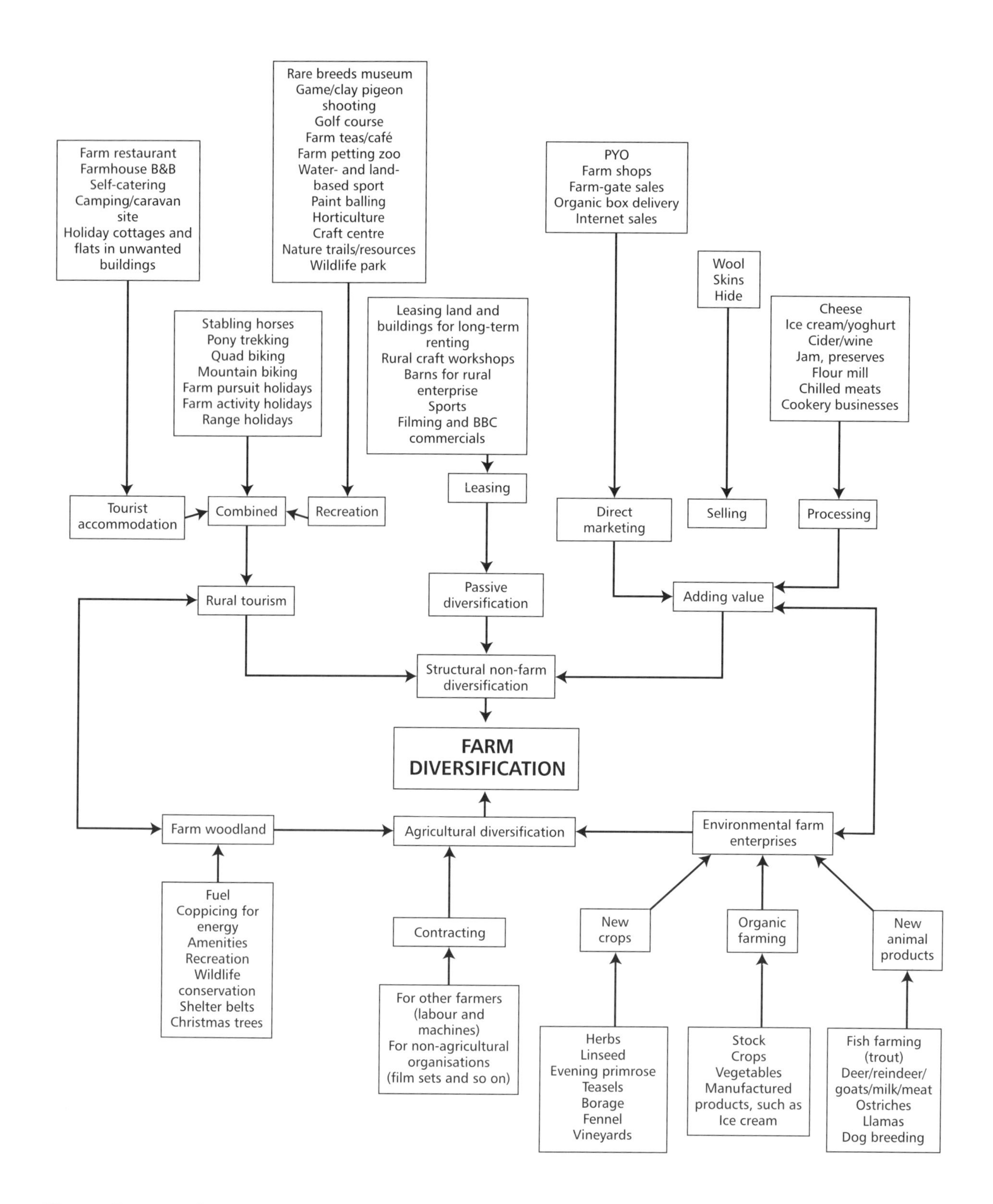

Figure 5.3 Diversify or die

Case study: Some examples of farm diversification

Waddington Grange, Lincolnshire

- Produces prepared dinner party ingredients for either a complete menu or a single course.
- Most of the ingredients are locally produced.
- The business is run by the wives of two farmers, but involves other farms in the supply of ingredients.

Pentrefoelas Farm, North Wales

- Embroiders names and motifs on towels and flannels.
- The business was started by the farmer's wife, with a £5000 loan from the Welsh Development Agency.
- It now employs nine farmers' wives and involves four embroidery machines.
- A shop has been set up to sell embroidered products and local crafts.

Wild Farm, Northumberland

- Offers a weekend-on-the-farm package.
- Besides full accommodation, the package includes farms walks, visits to other farms in the vicinity, horse-riding and fishing.

Halzephron Farm, Cornwall

- Consists of a 1.5 ha plot growing a range of basic herbs.
- The farm shop sells potted herbs and also a range of marinades, sauces and dressings made on the farm from the herbs.
- There is now a thriving mail order business.
- Annual turnover has grown to around £1 million and a staff of 16 is employed all year round.

Review

4 To what extent do you think that organic farming is the way ahead, particularly in agriculturally marginal areas? Re-read the first two case studies in this section. A useful research site for information on organic farming is:

http://www.soilassociation.org/

5 Referring to the four case studies opposite, explain how they benefit the farmer and local rural communities.

SECTION C

Producing more in the South

Whilst there was a doubling of food production in the South during the period from 1960 to 1990, the overall achievement masks substantial regional variations. Only in the Far East has food production consistently outpaced population growth. In Sub-Saharan Africa, the production of food per head fluctuated, but has actually fallen in some recent years. This fall has disproportionately hit the poorest countries and the poorest people within the LEDCs.

So what are the key factors in this prevailingly unfavourable food–people balance in the South? The reality is that the South faces a number of tremendous challenges. It has a population of 4 billion – and that is still rising. It is undergoing rapid urbanisation, so that today nearly 2 billion people in LEDC cities have to be fed; they are not producing their own food. Then there is the rapid pace and high cost of technological change.

Copying high-tech Western methods is not the automatic solution to the South's food problem.

There is also a range of issues involved that lie beyond actually producing more food:

- cutting down on storage losses, as well as losses between producer and consumer
- improving the efficiency of transport and other infrastructures
- correcting the inequitable distribution of land between rich and poor
- dealing with the growing unpredictability of harsh physical environments
- coping with the huge numbers of people living below the poverty line in both rural and urban areas, and who cannot afford the higher prices of imported food
- counteracting the increasing emphasis on cash crops grown at the expense of subsistence crops
- facing an uncertain future, as the impact of the Green Revolution levels off.

So far, the increase in agricultural production has resulted from two main developments:

- **extensification** – increasing the amount of agricultural land
- **intensification** – producing more from existing farmland.

Figure 5.4 Sources of increased food output during the second half of the 20th century

	Percentage contribution to increased food output	
Decade	**Extensification**	**Intensification**
1950s	82	18
1960s	25	75
1970s	16	84
1980s	2	98
1990s	3	97

Expert assessments of the relative contributions made by each of these developments are shown in **5.4**.

Why did the balance shift so much between the two sources? The prime reason was the intensifying effects of the Green Revolution, as they gradually diffused across most parts of the South. A secondary reason was that the expansion of agricultural land has only gone on until the most suitable land has been taken in. In the 1950s, for example, there was still 'uncultivated' land in much of Africa, Latin America and South-East Asia. As quality land was used up, so farmers had to turn to more marginal land. Farming at or beyond the margin increases the risk factor, especially as many farmers have little capital to purchase even intermediate technology (a level of technology that matches the needs and skills of LEDC peoples).

Figure 5.5 The gains and losses of extensification

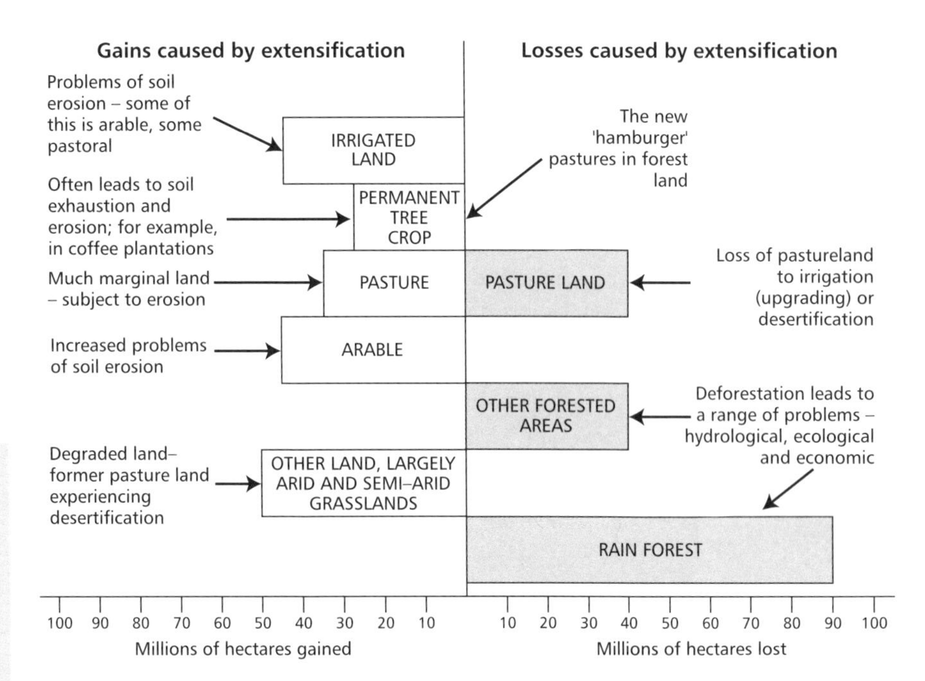

Review

6 Distinguish between **extensification** and **intensification**. How does the meaning of the former term differ between the MEDCs and the LEDCs?

7 Explain why farming beyond the margin increases the risk factor.

A number of environmental problems caused by extensification are highlighted in **5.5**. It shows there have been major changes in land use. Whilst there has been much successful colonisation of new land, especially in the rainforest basins of the Andes, there are numerous less successful examples, including the colonisation of the Brazilian rainforest in Rondonia associated with the building the Trans-Amazon Highway, or the Mahaveli project in Sri Lanka (see page 73).

SECTION D

The Green Revolution

Intensification of production per hectare is particularly linked to the basic food crops, especially the cereals that form the staple diet of most of the South. During the period from 1970 to 1990, there were spectacular increases in the production of nearly all cereals, particularly wheat, rice and maize, in nearly all regions.

Intensification can result from a number of biochemical, mechanical and social changes (**5.6**). During the period from 1950 to 2000, intensification was above all associated with the Green Revolution. This refers to the application of modern Western-style high-tech farming to LEDCs. It was particularly associated with the development in the 1960s of new hybrid varieties of wheat and maize, initially in Mexico. Later, new varieties of rice were developed in the Philippines, in particular the so called Miracle Rice (IR-8), which increased yields sixfold in its first harvest. In a wider sense, the Green Revolution involves a package of technology (including fertilisers, pesticides, herbicides, water control and mechanised farming equipment) needed to support the high-yielding varieties of seed (HYVs). The web of intensification in **5.6** shows the many facets of change associated with intensification. Clearly, the scale of the changes was such as to merit the term 'revolution'.

Subsequently, plant breeding moved on from cereals to develop improved strains of potatoes in Latin America. However, progress has been slow with other root crops such as yams. Animal breeding improved the yields of milk and meat; for example, from cross-breed cattle such as the Santa Gertrudis, which are now widely used in tropical Africa.

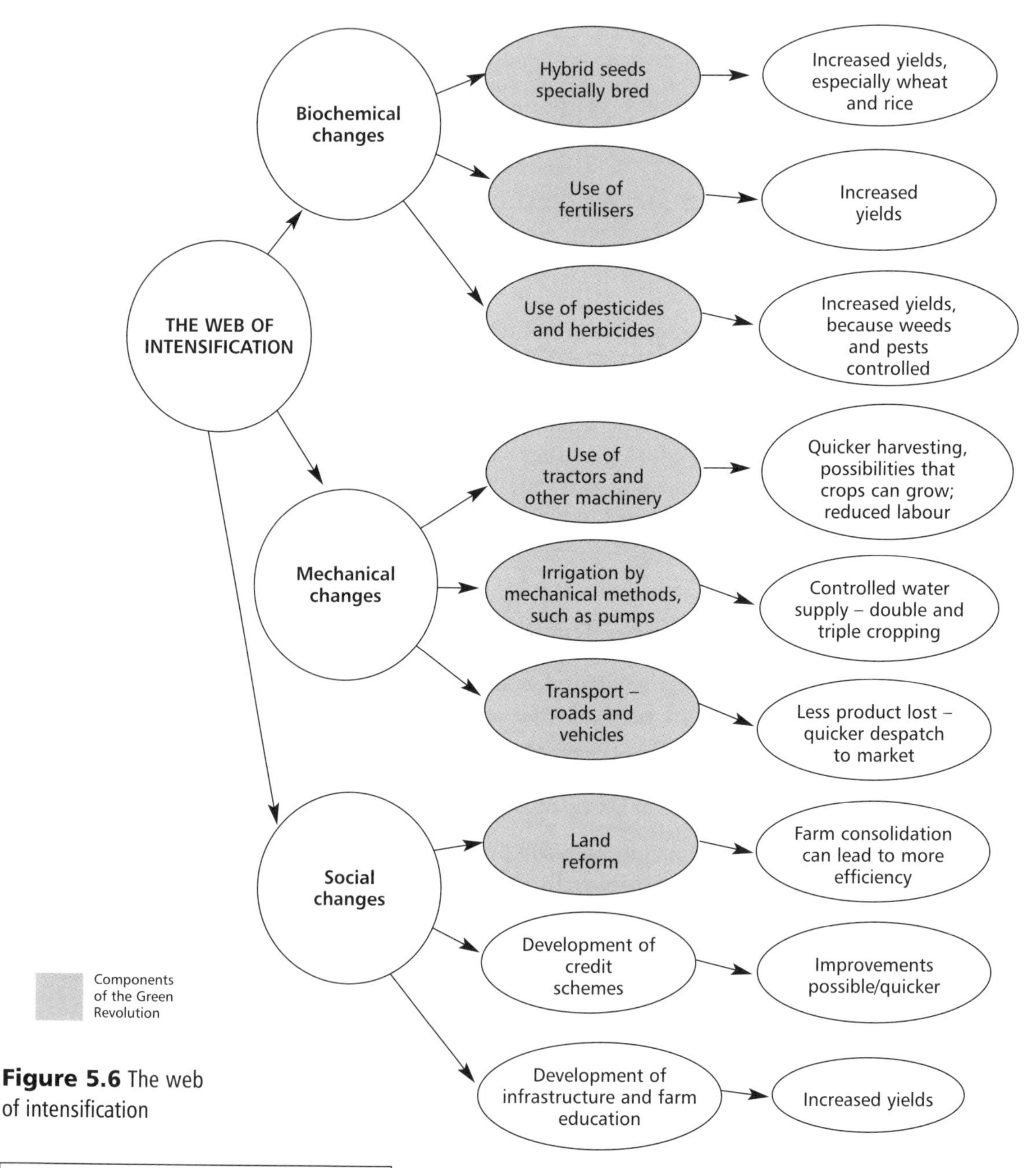

Figure 5.6 The web of intensification

It is fashionable to develop a critique of the Green Revolution. The first thrust of this must be that, whilst it undoubtedly increased food security enormously in South and South-East Asia, with countries such as India and Thailand actually becoming net exporters of grain, it did not increase food security in many other areas. Even in India, although grain is being exported, there still is no absolute food security for all. Indeed, in many parts of Latin America and Africa, the key to food security may well lie in the innovative, yet traditional methods of millions of peasant farmers (see page 64).

Figure 5.7 The advantages and disadvantages of HYVs

Advantages	Disadvantages
■ They give up to four times the yield. ■ Some varieties are more responsive to fertilisers and give higher yields per unit of fertiliser. ■ Many varieties are short and stalky and do not topple over when ripe. ■ Many varieties are not sensitive to day length and can be grown in a wide variety of environments. ■ They have a shorter growing period, and this sometimes allows double or even triple cropping.	■ Some varieties have poor milling qualities and can taste inferior. ■ They require heavy doses of chemical fertilisers. ■ They provide less straw for thatching and for use on the farm. ■ They are less adapted to drought and require more sophisticated water control from irrigation schemes. ■ Many varieties are more susceptible to pests and diseases, and require large applications of pesticides and fungicides. ■ New seeds must be bought each year to ensure the purity of the strain. ■ They need more careful weed control.

The introduction of new HYV varieties has not been problem-free (**5.7**). HYVs only achieve high yields under optimum conditions, which means careful water control, heavy application of fertilisers and frequent use of pesticides. HYV wheat, rice and maize have been much more successful than millet and sorghum. Whilst the diffusion of the Green Revolution was rapid in the 1970s and 1980s in favourable areas such as the Punjab, which are well endowed with water, fertile land and nearby markets, even there it was not the means by which poor farmers could secure their subsistence and also develop surplus harvests. Whilst there was some trickle-down to smaller farmers, they struggled with the high cost of inputs, and mechanisation meant that many landless farm labourers became unemployed. The Green Revolution has meant little to less favoured upland areas remote from urban markets, such as the Bolivian Alti Plano.

Another major concern is that an energy-consuming, high-tech agriculture increases dependency on external supplies of fuel, fertiliser, pesticides and seeds, all of which are supplied by TNCs. This places the Third World farmer at the mercy of those companies and global fluctuations in the price of inputs.

Selective plant breeding has been going on since people first planted crops – they pick the best seeds for next year's sowing. But harvest yields have stayed much the same. Some 20 years ago, however, researchers found new varieties of high-yielding seeds – wheat, maize and rice – which could double harvests. By 1985, Green Revolution seeds were being sown on half the area devoted to rice and wheat in the developing world

+ The world is better off because there is far more food available from bigger harvests than 20 years ago.

− Bigger granaries storing more grain are not enough. Hungry people need to be eating more. Food in storage is not being distributed to the needy.

+ Significant and populous countries such as India, Indonesia and Thailand have become self-sufficient in basic foodstuffs. They are no longer dependent on North American and European food aid.

− But implementing Green Revolution policies has increased dependency on imported seeds, fertilisers, pesticides and farm machinery.

+ Former food importers such as India and Thailand now export grain, earning useful foreign exchange. Initially the taste of rice products was a problem.

− Imports of seeds, petrochemical fertilisers and fuel for machinery all cost valuable foreign exchange, and gives control to agrochemical multinationals.

+ New farming methods with irrigation can bring all-year-round employment. No longer do workers have to be laid off in the dry season. Double cropping is common.

− Often, agricultural profits are invested in tractors which reduce employment.

+ With bigger harvests, farmers earn more and the price of food stays constant or even becomes cheaper in the market-place.

− For the small farmers who cannot afford the Green Revolution seeds and other technology, the lower prices for their harvests mean real hardship. Often they have to sell off to the big landowners, bringing an increasing gap between rich and poor.

+ The environmental impact of improverished rural people – who cut down trees – is lessened by the urban migration.

− New farming methods can bring an increase in water-borne diseases (with irrigation), the development of 'super-pests' (resistant to insecticides) and desertification (through the salinisation of waterlogged fields). Increased fertiliser use leads to eutrophication. Biodiversity is lost as native breeds are replaced by HYVs.

+ −
ASSESSMENT

Increasing food production does not stem hunger on its own. It is possible to have both more food and more hunger. 'If the poor don't have money to buy food', a World Bank report said, 'increased production won't help them.' Nevertheless, increasing harvests by 30% is the equivalent of discovering 30% more farming land; and as the population is increasing, that is a welcome relief.

Figure 5.8 The benefits and costs of the Green Revolution

Review

8 Write a short account highlighting the main features shown by the data in **5.6**. Are the changes shown the outcome of extensification or intensification? Justify your answer.

9 Using **5.7** and **5.8** and your own research, write a critique of the benefits and costs of the Green Revolution.

SECTION E

'Bottom up' initiatives

A third way of raising food production in those countries most in need is by transforming peasant agriculture from within. This has to be the long-term strategy for guaranteeing food security in the South, especially in the problem region of Sub-Saharan Africa. The recent history books are full of grandiose projects that have sought to raise food production in LEDCs by translocating Western-style technology to fragile tropical environments. Most of these 'top down' ventures have proved to be 'megaflops' which, if anything, have exacerbated the food-security gap between rich and poor in rural areas. The Mahaveli irrigation project in Sri Lanka is one of many that underline the point.

Case study: the Mahaveli irrigation project

Project goal – launched in 1978, the irrigation scheme was designed to help modernise agriculture and manufacturing in Sri Lanka. It involved building three dams along Sri Lanka's 300 km Mahaveli River and constructing power stations to triple the country's energy-generating capacity. It also involved irrigating 120 000 ha of new land and the resettling of 1.5 million people.

Cost and donors – the project was financed primarily by the UK, Sweden and West Germany, each giving around $100 million. Much of that financing came in grants, but large sums also had to be borrowed on 'soft' commercial terms to make up the difference. The project has been plagued with financial problems and the budget has nearly doubled to $14 billion. To meet the spiralling costs, the Sri Lankan government has raised taxes and cut food subsidies.

Winners – most construction contracts were awarded to companies in the donor countries, who employ expensive foreign personnel and machinery. Most of the newly irrigated land was leased to foreign companies using expensive imported machinery and very little Sri Lanka labour.

Losers – small farmers are being forced off their land as thousands of hectares are lost to make way for dams and reservoirs. Nearly 1.5 million people have been forced to resettle on small plots of forested land, 90 km away from their homelands. The meagre compensation for lost crops and the cost of land-clearing and new houses has already been cut by a third by the financially stretched government. The scheme threatens the habitat of many rare animals (the Indian elephant and the leopard) as well as 35 plant species native to the river valley. Already, two of the three dams have had to undergo frequent repairs.

There are many constraints facing peasant farmers. These include:

- the small size of their holdings (often only 2 ha, or the size of a football pitch)
- unpredictable weather, which can spell disaster
- the problems of bad health (AIDS in Sub-Saharan Africa) within a system that is reliant on labour inputs from the family and community
- fluctuating conditions, which require the peasant to go to an extortionate money lender in times of hardship.

However, the peasant farmers in many LEDCs have had a long-standing tradition of working with the environment to conserve soil or water and using internal resources such as human labour and animal power. This might involve terracing slopes or maintaining soil fertility by fallowing or by the heavy input of green or animal manure and human waste. The careful yet innovative approach of many Chinese peasant farmers (see page 98) is testimony to the widespread existence of sustainable yet productive farming systems.

Any successful attempt to increase agricultural productivity is likely to be characterised by some, if not all, of the following features:

- the identification of alternative technological solutions using local labour and raw materials, and utilising local expertise
- the concentration on small-scale community enterprises
- the encouragement of self-respect and self-reliance, by encouraging the indigenous society to find local solutions to problems of work and community
- the adoption of a low-level approach to advice, always seeking to train local people to provide advice
- the minimal use of pesticides and artificial fertilisers.

The charities providing rural aid, now known collectively as the non-governmental organisations (NGOs), have raised extensive funds as a result of high-profile, media-supported events such as Children in Need and Band Aid, or sponsored mass events such as the London Marathon. They now direct their efforts towards the poorest farmers by means of small-scale projects, of which agricultural development is an integrated feature. Many of these projects are 'bottom up' from the village roots, using appropriate technology to work with the local environment. It is difficult to assess the overall impact of these schemes, because many are very localised. Their cumulative effect, even in countries such as India, may not be nearly as important as the Green Revolution in terms of yields raised. These small-scale projects require an enormous investment of time and energy, and the diffusion of good practice from one rural village to another does not always happen. But they do represent a recognition of the value of locally developed, sustainable, low-tech solutions. They are beginning to change our thinking about foreign investment and aid (see **Chapter 7**). The websites of WaterAid, Farm Africa, Action Aid, Cafod and the Intermediate Technology Group give details of an enormous range of successful rural development projects.

Case study: 'Stemming the flow' developments in Burkina Faso

Plan International works with 35 000 families in dozens of villages in this small West African state. It employs bright young people to go into the villages to listen to the people and help them identify their needs. Plan International works with Tree Aid (another NGO), teaching about soil conservation techniques, and encouraging village projects to improve the basic infrastructure.

One major project has involved showing villagers how to save up to one-third of their scarce fuelwood by using efficient stoves. The outcome of this has been that hours formerly spent collecting fuelwood are now available for farm work. Food output has increased. Another project has involved tapping a nearby lake to provide irrigation for growing vegetables. These two developments have begun to transform village economies.

Tamassogo, a typical village, now has a five-year development plan that includes building a small mill and a youth centre, setting up a vaccination scheme for livestock and a cereal seed-bank. They have recently submitted a bid for a grant of £5000 to pay for sinking a new borehole.

This example of self-organisation and new commitment is encouraging, particularly because its one of its results has been to stop the outflow of young villagers to seek work in the Ivory Coast.

Review

10 Compare the relative merits of 'top down' and 'bottom up' approaches in helping to raise food production in the South.

SECTION F

North–South transfers

Figure **5.9** identifies some of the transfers taking place between North and South. You would think that it would be straightforward to transfer the food surpluses of the North to the food deficit zones of the South, but the relationship between the North and the South is both complex and dynamic. As we saw in the first part of this chapter, it is unlikely that the North will ever again build up the huge mountains of surplus food. We also saw that the South has an increasing ability to produce more food but not necessarily to feed its people. The South is a very heterogeneous area, encompassing middle-income, less-developed and least-developed countries. It might be called 'a three-speed zone'. The case study below shows there are enormous difficulties in organising this, even within one country.

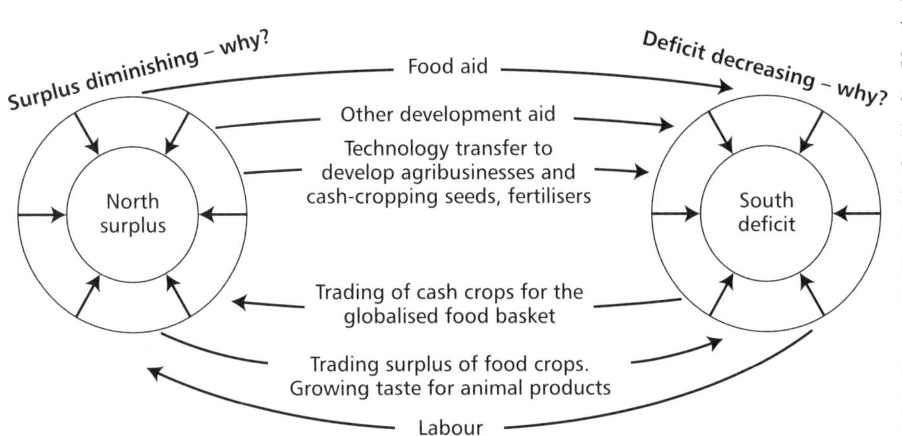

Figure 5.9 Some transfers between North and South

Case study: Problems of plenty – a new vision for India

The Punjab, India's breadbasket, is overflowing with food. Every silo, every cubic metre of storage space is filled, and bags of wheat and rice are being piled up in open lots around the state. 'Two months from now another crop of rice will be harvested', says a local official, 'and we have no space to store it'.

India is not only producing enough to feed its near 1 billion people, but it has also amassed a two-year stockpile of grain, and has even tried selling some of it abroad. India is just beginning to face the problems of plenty, and plenty of problems there are too. Conflicting government policies encourage overproduction by farmers, but discourage its distribution to the poor. A crumbling infrastructure makes it difficult to move food around the country to where it is needed, or even to ship it abroad. It is ironic that just as India has moved into the top tier of world food producers, perhaps as many as 300 million Indians remain malnourished.

The problems, like the surplus, are of the government's own making. Since 1965 it has agreed to pay farmers a minimum support price for their crops. They are free to sell to any buyer, but in practice the government buys 70–90 per cent of all the crops that come on the market. In recent years, it has been raising support prices to build up the nation's food reserves in the event of failed monsoons or other disasters.

One attractive option for a struggling economy with a grain surplus would be to export the overflow, but India cannot exploit that opportunity because of infrastructure problems. The Punjab, landlocked in India's far north, is linked to the rest of the country by manifestly inadequate transport links. When food does reach a port city, it faces a curious problem. The equipment is there to unload millions of tons of grain from incoming cargo ships (a relict of former aid programmes), but machinery for loading food into departing cargo ships is lacking. The government has decided to sell more than 4 millions tons of grain abroad, but in a country where so many people are still undernourished, exporting is politically difficult. Moreover, by raising the support price paid for grain, India has come close to pricing itself out of the international market.

So why not just give surplus food to the hungry? A few of India's 25 states have come close to doing this, with food-for-work schemes. But an estimated quarter of the nation's poorest poor live in remote areas or are nomadic. They have little opportunity to take advantage of such help, or even to shop at the nation's network of more than 420 000 subsidised Fair Price Shops. These shops offer grain and other foodstuffs at below-market prices. However, faced with budget problems and a strong

Review

11 Write short explanatory notes about each of the transfers shown in **5.9**.

12 Examine the factors that make it difficult to transfer food surpluses at a global and regional scale.

13 Explain how it is that India has a food surplus, but not all of its people are adequately fed.

farmers' lobby, the government has tried to increase revenue by hiking prices in those shops. Rice and wheat now sell for 20 cents and 13 cents per kg, almost double 1990 prices and nearly as much as the grains fetch in unsubsidised stores. As prices have increased, buying has decreased. Fair Price officials say the poor are now buying more grain on the open market or switching to cheaper items such as potatoes – or simply eating less.

It looks as if India's food surplus may not endure. Increased production is exhausting the soil and water in some areas. Because of irrigation, the Punjab water table is dropping 60 cm a year. In the 1960s, few would have expected that India could ever raise enough food to feed its population, but the country has done that and more. The next challenge is to use the abundance to the advantage of all its people.

SECTION G

Food aid, trade and agribusiness

Giving aid in the form of food is perhaps the most obvious transfer to make between North and South. It is particularly appropriate in emergency situations where severe food shortages are created by natural disasters or wars. Since 1990, the amount of food aid has declined by over one half, the cutback being most marked in programme, as opposed to emergency, aid. So why has food aid dropped so much?

- World Trade Organisation agreements not to subsidise agricultural production so heavily have cut grain surpluses in MEDCs.
- Food aid has to be bought with the 'aid dollar' and then given away. Aid agencies are now looking very carefully at whether food aid is such a good idea, and whether it provides as high a return in the fight against poverty as other possible forms of aid. Clearly, there is a need to respond rapidly to emergency situations for humanitarian reasons, but food aid cannot always be provided rapidly enough in substantial amounts, because of the inherent delays in the system of fund-raising, 'shipping' and delivering it to where it is needed. Sometimes the flows, once started, are so strong that the countries do not have the infrastructure to absorb them.
- Medium-term food aid programmes can depress food prices in local markets and actually act as a disincentive to local farmers to produce crops – hence the recent move towards 'food-for-work' schemes, in which people work on recovery programmes in return for guaranteed food.
- Food aid sometimes comes in inappropriate forms such as wheat and powdered milk. Wheat is not a staple of many countries of the South (rice, millet and maize are more usual) and this sometimes requires new cooking skills to bake a new type of bread. Equally, powdered milk has to be mixed with clean water to be used. The importation of new

types of grain can change taste preferences, forcing farmers to grow a crop that is inappropriate or unsuited to the local environment, or adding to the growing import bill. On the other hand, food aid can save lives in the short term, especially if there is better monitoring to anticipate climate problems or signs of political and social stress.

The likely trend is for emergency aid to continue, but for programme aid to be replaced by other measures to tackle the basic cause of famine. Whilst LEDC agricultural systems are increasingly developing the ability to feed their peoples, it is a moving target. With continuing high fertility, population growth rates remain high at between 2 to 3 per cent per annum. Then there is the added problem of feeding the growing numbers of urban poor resulting from massive rural–urban migration. A number of trade transfer options exist, such as importing food in return for primary commodities, manufactured goods or services.

Review

14 Use the FAO website and the *State of Food Handbook* to record the pattern of food aid in a recent year.

15 Draw up a table to show the costs and benefits of food aid.

Trade

Agriculture has long been the mainstay of South–North trade flows (**5.9**). Between 1970 and 1995, exports of foodstuffs from LEDCs increased more than sevenfold and accounted for roughly one-third of global trade in foodstuffs. However, commodity markets are highly volatile, and there has been a long run of decline in the real prices of many commodities, such as coffee and cocoa. This has had very serious consequences for countries in the South, the majority of which rely on one or two primary commodities for over 50 per cent of their exports.

It appears that LEDCs have had a raw deal when it has come to trading in the global market-place. The terms of trade have all too often been stacked against them. There is the argument (not widely accepted) that if these countries can earn more for their exports, then they could then afford to import more food. Happily, examples of fair trade goods are growing, from ethnically traded tea, coffee and cocoa to rural crafts. Another approach is to produce higher-value, organically grown produce, such as cotton in Egypt and fruit in Chile.

Agribusiness and the TNCs

A major issue concerns agribusiness. As we saw in **Chapter 3**, this involves the application of modern methods of organisation and technology to create a vertically integrated enterprise, handling agricultural products from the farm to the consumer. Agribusiness is being increasingly dominated by TNCs. A relatively small number of TNCs – such as Unilever, Nestlé, Tate & Lyle, Lonrho, Hoeschst, BAT and Monsanto – have an enormous influence, that extends to every aspect of agriculture: the ownership of land, agrochemical factories, shipping companies, marketing organisations, research facilities and training institutes, and even the retail outlets where the produce is sold to the consumer. They are also the big

Review

16 Debate or discuss the claim that 'the arrival of agribusiness and the TNC is good news for the LEDC'.

17 Explain why an increasing demand for meat and animal products in LEDCs is a threat to food security.

names in GM crops. As large-scale enterprises, often with a turnover larger than that of the LEDCs in which they operate, TNCs can dictate where, what and when crops are grown, control the quality and price, and even shape consumer tastes.

The agribusiness model has a number of features that give cause for concern. This is perhaps best typified by the domination of the banana market by the giant US TNC brands, such as Geest and Chiquita, at the expense of traditional Windward Islands producers. The TNCs adopt a high-tech approach, using intensive farming methods to maximise yields. This can lead to serious environmental degradation. Industrial agriculture uses expensive inputs of fertilisers, pesticides and herbicides, and requires mechanisation. It is not energy efficient and relies on costly foreign inputs. This is a particular problem in the latest low-risk (for the company) way of working, known as the **outworker system**. By this, the TNC contracts farmers in LEDCs to produce a fixed amount of produce at a certain price and by a given date. This means that many of these farmers can lose their livelihoods if they fail to deliver the quantity or quality required, or if the price paid for the product barely covers the costs.

Initially, many LEDC governments were very much in favour of the establishment of prestigious agribusinesses. They are a particular feature of agriculture in Zimbabwe, Thailand, Kenya and the Philippines, and there is no doubt that they make a significant contribution to the global food market. You only need to look at your local supermarket to see this. Undoubtedly, the export of these products does make a significant contribution to the LEDC trade balance, but so much of the profits go to the TNCs.

A further issue is that many of the agribusinesses operate on the best quality land, and use some of the most innovative farmers as their outworkers. The result is that the production of food crops loses both land and expertise to the growing of cash crops. Whilst some agribusiness products are for domestic consumption, the majority are too highly priced and are, therefore, exported. This change ultimately means that the LEDCs become dependent on food imports from MEDCs. So, in conclusion, the technology transfer is not all good news for feeding the growing populations of the South.

As the global food market develops, the new elite of the LEDCs acquire Westernised tastes, particularly in eating much greater amounts of animal protein. As the LEDC livestock industry grows in response, this again places more demand on food crops for feeding animals. This would suggest that it is absolutely vital that provision is made for growing more food crops in the South. It is therefore crucial that the countries of the South be allowed to subsidise their agriculture in order to develop self-sufficiency, which is the only real basis for food security. Perhaps you can now see that North–South transfers are not necessarily the answer. This is a vital point, which is underlined by the next two chapters.

Enquiry

1. Select a rural parish and – by means of a questionnaire survey if possible – find out what attempts have been made by local farmers to diversify their activities and output. Try to evaluate the achievements.
2. Investigate one of the following projects: the Polloreste (Brazil), the Gezira (Sudan) and the ALCOSA (Guatemala). Use the Mahaveli Project case study as a model, paying particular attention to the winners and losers.
3. Visit the websites of two of the following major charities: WaterAid, Farm Africa, Action Aid, CAFOD and the Intermediate Technology Group. Find out about some of their current rural development projects and compare their aims and methods.

CHAPTER 6

Securing a future by acting globally

SECTION A

Food security

A concern about food security is something that all of us as individuals share, and so too all responsible governments. It does not matter whether we are living in the UK, Uganda or Uruguay. Food is the good that keeps us all alive; it is the overriding human need, the very means of life. Lack of food is the ultimate exclusion. When people do not have food, they are excluded from what the rest of society is doing regularly – eating.

Figure 6.1 The main causes of food insecurity

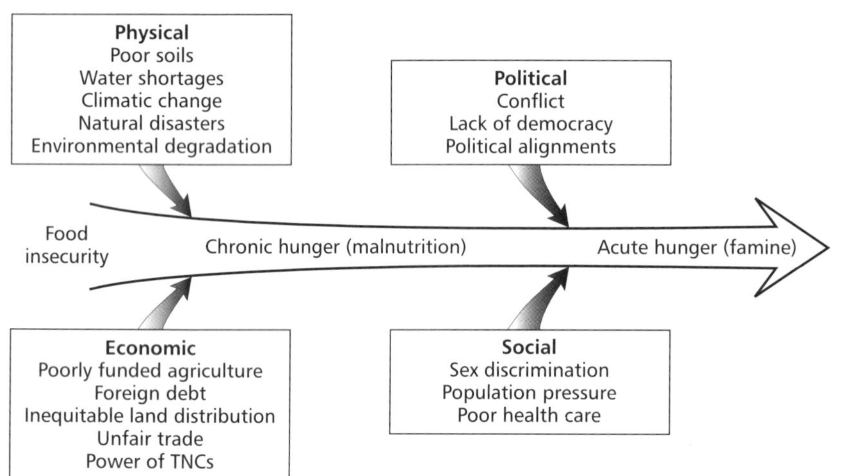

In this and the last chapter, the central theme is food security. The aim is to focus particularly on what might be done to ensure that a secure food supply becomes a basic human right possessed by all the global population and by all nations, rather than by just a privileged quarter. Figure **6.1** shows what many consider to be the main causes of food insecurity. Grouped under four headings, they may be seen as driving food insecurity from occasional shortages of food, through increasing degrees of chronic hunger (malnutrition) to acute hunger (famine).

In identifying the possible causes of food insecurity, **6.1** signals a range of action targets that need to be hit in order the rectify the situation. In exploring possible courses of action, the discussion touches on a number of topics that might be described as topical issues in global farming today. Possible actions may be classified under one of two headings. First, there are those which are more appropriately considered and applied at the global scale, such as reforming world trade and aid, checking global warming, as well as grappling with the highly controversial issue of GM crops. Second, there is a whole range of initiatives that are better taken at more 'local' levels, from national downwards. These include searching for sustainable modes of farming, exploiting alternative sources of food, land reform and revising traditional gender attitudes. They will be examined in the next chapter. For now, the focus is on acting globally.

Review

1 Study 6.1 and suggest how each of the factors contributes to food insecurity.

2 Can you think of any other factors that might be added to the four boxes?

Liberalising trade

There is no doubt that liberalising global trade (making it freer and fairer) would allow agricultural products from LEDCs better access to the lucrative MEDC markets. Equally, recent meetings of the World Trade Organisation have clearly indicated that there are powerful objectors to such a freeing of trade, particularly that in agricultural products and food.

Case study: The three-way split of the WTO

Meetings of the World Trade Organisation in Seattle (1999) and subsequently have shown the global community to be divided into three 'camps' on the issue of trade in agricultural products.

The European Union and Japan – this camp is keen to maintain subsidies on food exports and quotas on food imports. From the viewpoint of the other two camps, the Common Agricultural Policy (CAP) is the greatest single obstacle to the liberalisation of trade. The EU argues that the CAP is important in sustaining jobs and stimulating rural economies. It is also keen for more stringent controls on food safety and quality; the other two camps see this as a move against them.

The USA and the Cairns group (15 grain-exporting countries) – this camp is pressing for the abolition of all barriers to trade in foodstuffs and ending measures such as export subsidies. The USA has a problem in that its government has reduced support for farmers on the pretext that they would be able to sell more abroad. But such is the nature of US foodstuffs that few people abroad want to buy them. US-based TNCs are keen to be allowed to trade freely in GM crops and foods.

The LEDCs – having opened their doors to foreign foodstuffs, they claim that the other two camps have not done likewise. The are particularly unhappy about the EU continuing to dump its food surpluses on them, which has a depressing effect on their own agriculture. The USA has increasingly used its anti-dumping measures to stop the entry of LEDC foods.

Even if there were to be signficant improvements in the fairness and openness of world trade, would there be genuine benefits for food security in LEDCs? The answer is most likely 'no'. The explanation is painfully simple. An opening of MEDC markets to LEDC food would provide a tremendous boost to commercial agriculture, but at the same time it would be to the detriment of subsistence farming (see **Section E**). The outcome would be more landless peasant farmers. More farmers would be persuaded to go for cash crops that do not yield sufficient cash to buy the food that is no longer grown, but has to be imported. In short, the outcome may be more exports, but certainly less food. Imported food would be beyond the means of most small-scale farmers.

Everyone knows that we live in an age of **economic globalisation**. There is the widespread belief, particularly among LEDCs, that if they do not join in, they will somehow miss out. However, there is a growing body of evidence that points to globalisation being an inequitable process. It favours the strong rather than the weak. When it comes to food supply and security, it would seem that the more an LEDC can meet its own needs, the less exposed it is to processes that are working against rather than for it. The world may be moving towards what some see as the **global village**, but that village community will definitely not be a classless one. The case study of Cuba, isolated from much of the world, shows that a nation can do well by going it alone, and without fair and free trade.

Case study: Cuba goes it alone

Recent developments in Cuba provide a clear illustration of what can happen when a country trades less. A long-standing trade embargo imposed on Cuba by the USA, plus the collapse of the island's sugar exports to the former Soviet Union, led to a switch to farming based on low external inputs that has improved food security. Because it has been isolated from the chemicals and machinery necessary for modern intensive agriculture, it has been forced to turn to organic farming. It is claimed that the results achieved 'have overturned the myths about the inefficiency of organic farming'.

The most significant steps in the transformation include the following:

- oxen have been bred to replace tractors – their manure is used to 'feed' the soil
- a biological pest control programme has been developed, in which farmers produce biocontrol agents instead of pesticides to protect their crops
- crop rotation, green manuring and intercropping are now widespread practices
- urban farming has been encouraged – even in Havana, the capital city – and urban dwellers to take time out to work on farms.

Basically, two things have happened. Cuban farmers have remembered the old techiques, such as green manuring and intercropping, that their grandfathers used before the advent of modern fertilisers and pesticides. Cuban scientists have come up with affordable input-substituting bio-pesticides and bio-fertilisers. The best news of all is that food yields have increased and the country is beginning to enjoy an increasing measure of food security.

Review

3 Find out more about the WTO. When was it set up, what is its role and what are its achievements?

4 Go through the arguments in favour of an LEDC basing its food security on home production rather than trade.

5 What is meant by **green manuring** and **intercropping?**

Giving aid

The issue of aid was discussed in the previous chapter. All that is necessary now is to underline some aspects of it that might help food security. The critical bottom line is not so much the amount of aid as the form that it takes.

Figure 6.2 Breaking into the cycle of poverty and food insecurity

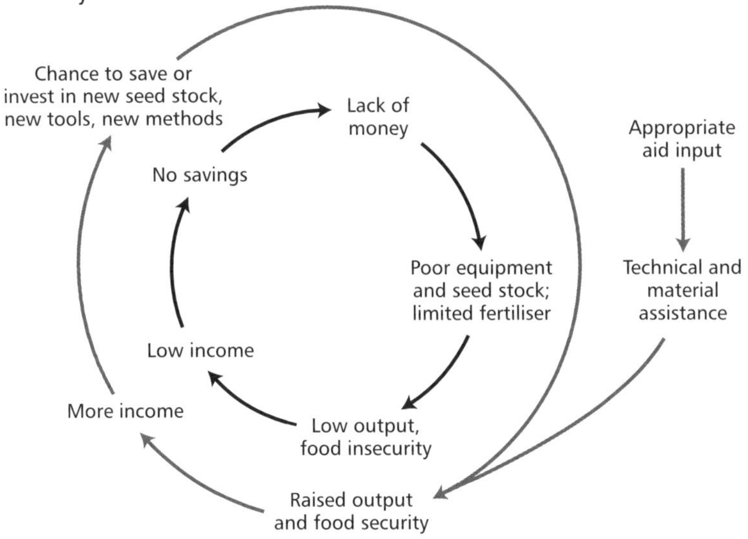

- Aid in the form of food is fine in disaster and emergency situations, such as during a famine. However, giving such aid during more 'normal' circumstances can only create an unwanted level of dependence. In all cases, it is better to be producing food rather than relying on handouts.
- Outside emergency situations, aid is best in the form of technical assistance. In the context of food security, this might involve helping local agriculture with seeds, stock, tools and basic instuction on the techniques of sustainable farming (see the Cuba case study and the permaculture case study in **Chapter 7**). It might also involve setting up village co-operatives, marketing groups and generally giving local communities the confidence to help themselves.
- The least desirable form of aid is the loan. Loans attract interest which has to be paid regularly, and loans ultimately have to be repaid. LEDCs are all too familiar with the vicious downward spiral created by increasing debt. On top of this, loan capital often seems to diminish, if not disappear, through the hands of unscrupulous government officials.
- Closely rivalling the loan as an undesirable form of aid is the so-called 'tied aid'. Agreements are reached between the donor and receiving nations that, in return for aid, the latter will purchase something made by the former. All too often, that 'something' has been armaments.

There are three ways of delivering aid:

- through **bilateral** agreements, made on a one-to-one basis between donor and receiver countries
- through **multilateral** channels, such as the various agencies of the United Nations and other governmental organisations
- through the **voluntary** sector of NGOs, such as Christian Aid, Oxfam and the Red Cross.

All three should be capable of delivering aid in an appropriate form. However, when it comes to delivering aid to where it really matters and breaking into the cycle of poverty, the record of the NGOs is probably best (**6.2**).

Review

6 What type of aid is most likely to benefit food security?

7 Why do you think that the voluntary organisations are rather better at delivering aid to where it really counts?

SECTION D

Pushing GM crops

During the 1990s, the Green Revolution began to give way to the Gene Revolution. The passage is proving to be highly controversial. No one doubts the ability of biotechnology to manipulate the genes of crops and livestock, but for what purpose? Supporters would answer that, by improving disease resistance, the food supply will not only increase, but become more secure. Opponents suspect that there is a hidden agenda, with TNCs seeking to raise profits by tightening their grip on agriculture and the food industry. At the moment, the debate centres on genetically modified (GM) crops and foods.

By early 2000, there were nearly 25 million ha of farmland growing GM crops, chiefly soya and maize, grown on large, capital-intensive farms in the USA, Canada, Mexico and Argentina. More than half the soya beans were grown from GM seeds produced by the TNC Monsanto (which has recently taken to trading under the name of Pharmacia). Most of the global output of GM crops so far has gone for animal feed. Although only a small proportion has been offered directly for human consumption, people are consuming GM crops indirectly through animals.

The GM food industry is dominated by five large companies – Monsanto, AstraZeneca, DuPont, Novartis and Aventis. In the USA, Monsanto controls over 85 per cent of the GM seed market. These TNCs need a large international trade in GM crops to pay off huge R&D costs and to realise their profit potential. They are being offered to the world as the essential scientific breakthrough needed to feed the world and reduce poverty and hunger in LEDCs (**6.3**). The trade has increased, but not as much as was hoped. Many countries, including all members of the EU, have refused to import GM foods and seeds.

Figure 6.3 Biotech's global takeover

Five major concerns have emerged:

- There is as yet no evidence that GM crops are capable of yielding more food per hectare of land. If this remains the case, then why promote them?

- GM crops are unlikely to benefit LEDC farmers. Not only is it expensive, but the GM technology threatens long-established and reasonably sustainable local agricultural systems. It gives TNCs monopoly control over seed stocks and therefore food production.
- There is concern about the impact on human health of consuming GM foods. There is no evidence as yet of any adverse impacts, but it is early days. It may take generations before it really shows.
- Likewise, possible environmental impacts need much more research. Field trials are being conducted in the greatest of secrecy, particularly in LEDCs where there are fewer official controls.
- When it comes to motives, no one can complain too loudly that TNCs are businesses. They are motivated by profit and are out to increase market share. It is when one looks more closely at how they intend to achieve those objectives that the concern starts.

Case study: Monsanto – a monster TNC

- Monsanto, based in St Louis (Missouri) is probably the largest producer and distributor of agro-chemicals in the world. But this is only one part of its huge business portfolio. It has recently been renamed Pharmacia.

- Monsanto has spent billions of dollars on agricultural biotechnology over the last five years. Genetically engineered products include Bollguard cotton, New Leaf potatoes and Yieldguard corn, all of which are designed to be resistant to specific insect pests.

- The corporation is strengthening its position in the market by buying up other biotechnology companies. These include seed companies, particularly those that have carried out R&D on cotton and maize. These acquisitions are a means of securing worldwide dominance for its genetically engineered products.

- Monsanto spent around half a billion dollars to develop the Roundup Ready soya beans, which contain foreign genes from a virus, a petunia and a bacterium. The beans are designed to be resistant to Monsanto's Roundup, the world's best selling herbicide, which accounts for 15 per cent of the company's sales and 40 per cent of its operating profit.

- Monsanto developed and produced the Agent Orange defoliant used during the Vietnam War, as well as a substantial proportion of the world's PCBs, a group of chemicals recognised as being so hazardous that the US Congress took the unique decision to ban their production.

- In the past, Monsanto has been guilty of discharging highly toxic wastes into the environment. It is now committed to reducing all releases and emissions to the 'ultimate goal of zero effect'. Perhaps it might be pointed out to the corporation that it is its products, more than its wastes, that threaten human health and environmental stability!

- Monsanto promote themselves as part of the answer to the world's food and environmental problems, claiming that 'sustainable agriculture is only possible with biotechnology and imaginative chemistry'. If you believe that, then read **Chapter 7**!

Three examples illustrate the last point about the ways of gaining market share:

- Monsanto's convenient 'twin' production of Roundup and Roundup Ready soya seeds has already been referred to in the Monsanto case study – a case of a double whammy!
- Even more sinister is the so-called 'terminator technology' being developed in the USA. Basically, this technology produces seeds that will only germinate for one season. Reproduction of viable seed is stopped and the second-generation seed is made sterile. Such technology does two things. It 'terminates' the age-old farming practice of saving seed from one growing season for use in the next. It guarantees the producer company sales of new seeds every year. Monsanto, who now own the technology, argue that it would benefit the world farming community. Farmers in all parts of the world would be able to share in the advantage of improved seed!
- The TNCs are plundering the genetic reources of LEDCs and turning them to their own profit. The countries and peoples from which the original plant material comes see few of the benefits – or cash – derived from its use. For example, a wild tomato variety taken from Peru in 1962 to improve the sweetness of tomatoes has contributed $8 million a year to the US tomato-processing industry. None of these profits have been shared with Peru. On top of this, the TNCs are taking out patents on parts and processes of plants, animals and even people. Is it right that a company can patent Nature?

Enough has been said here to make the point that the development of GM crops has little or nothing to offer to LEDCs by way of food security (**6.4**). Genetic engineering is not the only way, nor is it the best way to increase food production. GM crops are not relevant to the main reason why people go hungry, namely the lack of money to buy food or the lack of land on which to grow it.

Review

8 Find out more about the possible impacts of GM foods on human health, and of GM crops on the environment.

9 Write a short essay entitled 'Biotechnology – a wasted opportunity'.

	What is needed?	What is being done?	What could be done?
CROP PRODUCTION	Poor people's crops need conserving and improving to make them more pest-resistant, more nourishing and higher-yielding.	Instead of making crops pest-resistant, some companies are making them chemical-resistant, to increase sales of chemicals. In general, only major cash crops are being bred to yield more.	Traditional crop varieties could be conserved and new crops selectively bred for hardiness.
ANIMAL HUSBANDRY	The Third World needs to conserve its genetic diversity. Poor people need livestock which live longer and produce more.	Attention is directed at complete control over animal reproduction, developing uniform (but very vulnerable) breeds of animals, and veterinary drugs and equipment.	Vaccines and ways of diagnosing diseases could be developed. Cross-breeding could create healthier, more efficient livestock.
FOOD PROCESSING	Poor people need cheap, nourishing, non-perishable food, produced in a culturally and environmentally friendly way.	The use of natural raw materials is being reduced or they are being substituted by artificial ones, and products traditionally thought of as agricultural are being produced in factories.	Traditional biotechnological methods of food preservation, such as fermentation, could be further developed.
HEALTH CARE	Third World people need clean water, preventive health care, improved sanitation and nutrition most of all. Next come new vaccines for tropical diseases.	Tools for medical diagnosis (rather than treatment) are being developed, along with hormone production and drugs to prevent aging and cancer. Organ transplants and gene therapy are also top priority.	Biotechnology could provide better techniques for water testing and vaccine production.

Figure 6.4 What biotechnology could do for LEDCs, but does not

SECTION E

Growing for export

During the 1980s and 1990s, food-producing TNCs and agribusinesses moved into LEDCs on a large scale. This was part and parcel of the process of economic globalisation. The main aim was to produce food for export to the increasingly affluent and discerning consumer markets of the MEDCs. The idea of producing crops for export appealed to LEDC governments keen to earn foreign currency to pay off aid debts. It also appealed because it was seen as a way of becoming involved in the process of economic globalisation. Perhaps it would kick-start economic development.

Unfortunately, the experience of many LEDCs is that the export trade in food crops has been at the expense of food for local people. Other costs have included environmental degradation, a decline in workers' health, an unequal distribution of economic benefits and an eclipse of government by powerful TNCs.

Case study: Chile counts the cost

In 1980, Chile exported about the same amount of beans, an important staple, as it grew for home consumption. But by the early 1990s, the quantity exported was almost three times greater. Between 1989 and 1993 the area under basic food crops fell by nearly 30 per cent. Fruit, flowers and other crops destined for the export market had replaced beans, wheat and other staple foods. Large-scale fruit producers bought out small farmers who could not afford to invest in the new crops.

All of this has changed the face of the country's agriculture. It has also embittered many small farmers, who have become poorer and less food secure as a consequence.

Review

10 Draw up a table listing the costs and benefits to an LEDC of exporting food crops.

SECTION F

Protecting biodiversity

Despite 10 000 years of settled agriculture and the fact that 50 000 species of edible plants have been identified, the human race uses no more than 200 on a regular basis. In fact, a mere 15 species account for 90 per cent of the world's food supply, with just three of them – rice, maize and wheat – supplying nearly two-thirds.

Figure 6.5 An early example of genetic erosion in the USA

Vegetable	Total number of varieties in 1903	Percentage of varieties lost by 1983
Artichoke	34	94.1
Asparagus	46	97.8
Runner beans	14	92.9
Lima beans	96	91.7
Garden beans	578	94.5
Beets	288	94.1

In the circumstances, you might be forgiven for wondering why such a fuss is being made about the loss of biodiversity, or **ecocide** as it is sometimes called. The nub of the problem is that plant breeding to produce new varieties of familiar crops does one thing and requires another. It causes **genetic erosion** and a reduction of the **gene pool**; it needs genes from the wild to introduce new traits. The same applies to livestock breeding. In both cases, the process of genetic erosion leads to genetic uniformity and increased vulnerability to disease. The point made by **6.5** is that the process of genetic erosion has quite a long history; the worry is that it accelerated considerably during the 1990s. In the 1970s, genetic erosion left the US maize crop vulnerable to a blight that nearly wiped out all production. Resistance to that blight was eventually found in a gene in a wild variety of maize found in Ethiopia. In India, ten types of rice are grown in an area that once contained 30 000 different varieties: this has significantly increased the risk of crop failure. But the problem is not just India's. During the 20th century, about 75 per cent of the genetic diversity of agricultural crops was lost, thereby reducing global food security.

Gene jargon

biosafety The impact of genetic engineering on the environment and health.

biotechnology The industrial use of biological processes.

gene pool The very basis of biodiversity.

genetic engineering A technique used to transfer genes from one organism to another, or to change genetic material within an organism.

genetic erosion A reduction in the number of species or varieties as a result of genetic engineering.

GMO Any plant, animal, micro-organism or virus that has been genetically modified or engineered.

novel food A term used by the food industry to describe genetically engineered food.

pharming The production of medicinal products from genetically engineered plants and animals.

transgenic Refers to an animal that contains genes from another species.

Case study: Forgotten foods

One way of compensating for the narrowing gene base of popular crops is to broaden the range of plants cultivated. Cultivation of roots and tubers, such as cassava, yams and taro, which are important staples in many LEDCs, could easily be expanded. They are all nutritious foods, high in carbohydrates, calcium and vitamin C. Moreover, they are easily grown in the Tropics, even on poor soils.

Another possibility is to bring back some of the world's forgotten foods, foods that were consumed in the past, but lost out to modern agriculture. In a number of LEDCs, particularly in the Tropics, under-exploited traditional food plants are being rediscovered. Amaranto and quinoa were two grains cultivated by the Incas of Peru and the Aztecs of Mexico. Both have been rediscovered by agronomists. These two grains are versatile and nutritious foods, that contain more high-quality protein than most other commercial grains, including rice, maize and wheat. There is a growing commercial market for both plants. Health food companies in Europe and North America have already developed products based on both grains.

Review

11 Check that you understand why biodiversity and the gene pool are so important to food security.

Amaranto and quinoa are only two of thousands of species that could be rediscovered and cultivated in commercial quantities. If today's global population of 6 billion people is to gain food security, then it seems sensible to broaden greatly the global food base from the few varieties on which we presently rely. In this sense, therefore, our survival depends on the survival of biodiversity.

Cooling global warming

It is not the purpose of this book to go into the reasons why the international community cannot agree on a programme of measures and actions to reduce the human contribution to global warming. The aim here is simply to outline what a failure to cut carbon emissions will do for food security (**6.6**).

Variability in climate, and therefore in agricultural production, is already a key factor in food insecurity. Some areas of the world are particularly prone to such variability. These include the Sahel of Africa, north-east Brazil, Mexico and Central Asia. Global warming looks set to increase this climatic variability. Higher temperatures will bring about a stronger atmospheric circulation and a faster water cycle. Rainfall may become heavier, but it will also become more erratic. Erratic rainfall means more drought, more drought means more irrigation. Certainly, a rising demand for irrigation water is going to put still more pressure on scarce water resources. There will be more stress on fragile farming systems; livestock disease and heat stress will also increase.

Global warming particularly threatens low-lying coastal areas due to rising sea levels. Some island countries, such as the Maldives in the Indian Ocean and Kiribati in the Pacific, may disappear beneath the sea. Elsewhere, the seepage of salt water into groundwater will have damaging effects on crop cultivation, particularly that of rice. For the densely populated areas of Bangladesh, China, Egypt, Indonesia and Malaysia, the prospects are grim indeed. Sea-level rises could also affect marine aquaculture worldwide.

Figure 6.6 Global warming and agriculture: threats, impacts and responses

Threat	Impact on agriculture	Human response
Rising sea level	Loss of farmland to the sea and to the relocation of displaced people	Build better sea defences; careful relocation of displaced population
Saltwater seepage	Crop deterioration; deterioration of grazing	Either abandon areas or biotechnologists to come up with salt-tolerant varieties
Rising temperatures	Favourable impacts in some areas, unfavourable ones in others	Take marginal land out of production
Increase in severe weather	Damage to crops and to farmers' profits	Plant windbreaks; improve drainage or irrigation
Extinction of plants and animals	None immediately, but genetic erosion may have consequences for future genetic engineering	Preserve as much of the gene pool as possible
New pests and diseases	Could be highly damaging to both crops and livestock	Another opportunity for the biotechnologists to come to the rescue

For subsistence farmers, global warming promises lower yields and a lessening of food security. Few of those farmers have the resources to make adjustments to the changing climatic scenario before the forecasted disaster actually happens. The current bickering between governments (mainly of MEDCs) about measures to combat global warming hardly bodes well for LEDC food security.

The conclusions to be drawn from this discussion are left until the end of the next chapter. By then, initiatives of a more 'local' nature will have been examined. We should then be in a better position to evaluate the relative merits of acting globally and acting locally. Does one promise better food security than the other? Or should the strategy be to 'pick and mix' the best actions from both?

Review

12 Check that you understand the main causes of global warming.

13 Explain why global warming threatens double trouble for the world's coastal regions.

Enquiry

1 Visit the website of one of the voluntary aid organisations and find out how they are helping to promote sustainable agriculture.

2 Choose one of the other leading agro-chemical TNCs (AstraZeneca, DuPont, Novartis or Aventis). Visit their websites, and find out about their products and how they promote them.

CHAPTER

7 Securing a future by acting locally

In this final chapter, the spotlight falls on a range of actions that might be taken more at a national level to achieve better food security. The actions are essentially of a grass roots nature. As will be illustrated by the case studies, some are quite simple and local in character, whilst others probably require quite fundamental changes in a nation's society.

SECTION A

Controlling population

It is a widely held belief that population numbers are a major factor in food security. It is tempting to think that if population growth were properly controlled, then hunger would disappear within a generation. That may be true for some LEDCs, but the global food problem is not a shortfall in quantity but, rather, one of poor distribution. Indeed, there is an increasing number of countries – admittedly most are MEDCs – where population numbers have been so controlled as to create a series of problems, some of which are to do with food (that is, producing too much for a declining demand) whilst some are unrelated (for example, providing adequate welfare services for 'wrinkling' populations).

On this whole issue of population numbers and their control, there is a growing body of evidence that supports the Boserup rather than the neo-Malthusian viewpoint. At a global level, food output has kept just ahead of population growth. The world produces more food per capita than ever before, but there was a worrying downturn in the 1990s so far as cereals are concerned. There is no relationship between the prevalence of hunger in a country and its population densities. For every densely populated and hungry nation such as Bangladesh or Haiti, there is a sparsely populated and hungry nation such as Brazil or Indonesia. Population growth is not the only pressure on land. In some LEDCs, expanding agribusinesses and TNCs are literally pushing small farmers off their land. They are also being squeezed out because the output of a small farm cannot compete with that of monocropped farms that are growing the latest high-yielding varieties. It is actions such as these that are driving them towards poverty and hunger. In other LEDCs, it is a shortage of labour that is threatening to cause food shortages. In a number of Sub-Saharan countries, farming populations are being decimated by AIDS, and this is reaching the point at which the ability to produce food is becoming seriously affected by a lack of able-bodied people.

Case study: The birth dearth

The case study on page 64 concluded that China's tough birth control programme had played some part in improving per capita food supply and reducing hunger in that giant of a country. However, there is an increasing number of instances in which a cutting of growth rates has been achieved without any national birth-control programme. Across Africa and Asia, millions of people are confounding demographers' predictions about rates of population growth by reducing family size. It used to be argued that without the strong imposition of birth control, child-bearing only declines when economic conditions improve. During the 1990s, however, a number of LEDCs disproved this. At one time, Bangladesh was held up to shame as having a breeding population that was out of control. In the last decade, the fertility rate fell from 6.2 to 3.4 children per women of child-bearing age. Despite acute poverty and widespread illiteracy, the fall in fertility was due largely to increased contraception. Fertility rates in many African countries, although still high, are also beginning to fall as the practice of birth control grows there as well. The fall in fertility is thought to be 'crisis-led' and to reflect the fact that when faced by increasing hardship, particularly food shortages and hunger, people choose to have smaller families. Nonethless, the decline in fertility in the LEDCs has a long way to go to match the situation in the MEDCs, where rates are now below the critical 2.1 replacement level. Here, possibly, is evidence to support the neo-Malthusian view.

Review

1 Why is population growth a national rather than a global problem?

2 Are you able to suggest an explanation for the downturn in per capita cereal production in the 1990s?

3 Does the idea of a crisis-led fall in fertility lend support to either the Boserup or the Malthusian view of population growth?

SECTION B

Going for gender equality

It is estimated that women farmers produce a large proportion of the world's food – between 80 and 90 per cent in Sub-Saharan Africa, 50–90 per cent in Asia and 30 per cent in Eastern Europe. As an FAO official has put it, 'There would be no food security without rural women.'

Women not only grow food, but in many countries they are responsible for processing and marketing. Despite all this, food insecurity for women in LEDCs is rather more serious than for men. In many countries, men receive a larger share of food than women. Thus hunger often strikes women first, and sets in motion a vicious circle of female malnutrition (**7.1**).

There are at least four other vital areas in which women are disadvantaged in so many 'traditional' societies:

- Although women work the land, they are precluded from owning it and do not have any sort of legal right to it. This means that they have great difficulty in obtaining credit facilities and other forms of help. This, in turn, tends to inhibit any form of agricultural improvement.

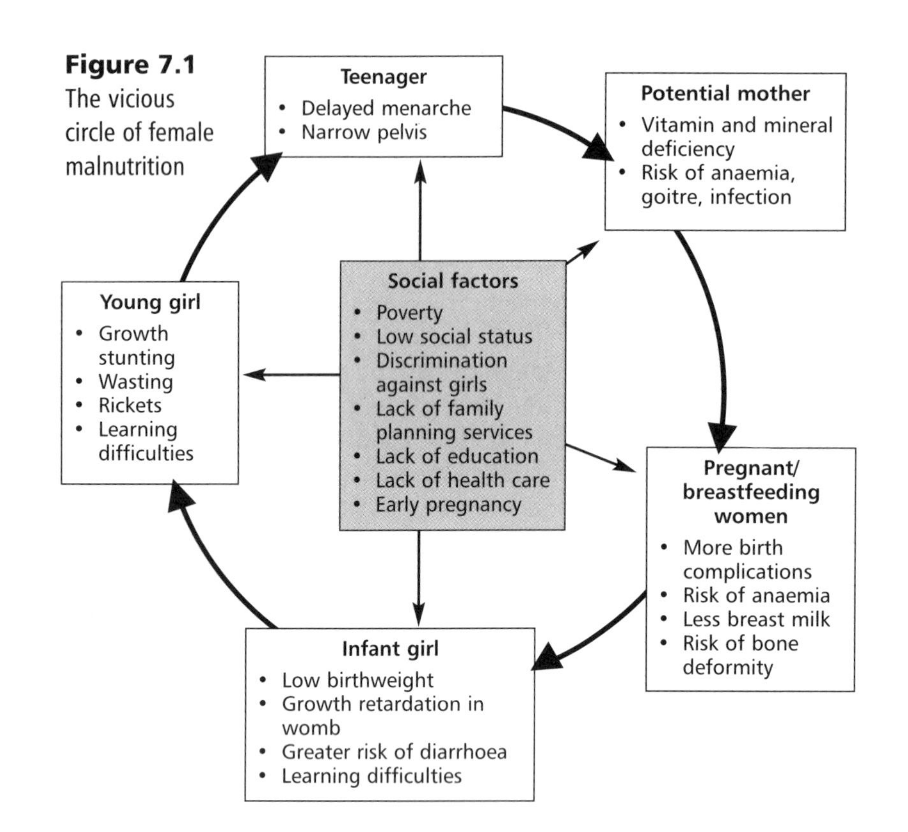

Figure 7.1 The vicious circle of female malnutrition

- Two-thirds of the world's illiterate people are female – fewer girls go to school than boys. The knock-on effect here is that women are often ill-informed as to their rights.
- Decision-making is male-dominated, and as a consequence the potential of half the population tends to be overlooked and unexploited, other than in the most menial of ways.
- There are still countries that deny women the right to vote.

What is very evident is that simply recognising and respecting the rights of women would do much to improve the practices and output of subsistence farming in many parts of the South. It seems a simple step to take and one that would ensure greater food security, but how many LEDC governments have the guts to do so?

Case study: An East African tale of two women farmers

Three weeks after her husband died, 28-year-old Maria Mathai found that she and her four daughters had no rights over the 2 ha farm that the family had worked on the slopes of Mount Kenya. When the title deed of the farm arrived, it was in the name of her late husband. As a widow with no son, she was not entitled to inherit the farm, despite the fact that she worked on it much more than her husband. She was asked by the chief of her husband's clan to move out of the village, and so join the growing volume of dispossessed women who have no option but to migrate to Kenya's towns and cities in search of work.

Like 85 per cent of Zambia's farmers, Joyce Kayaya works on a very small scale, but these small farmers produce something like 60 per cent of the country's supply of maize. She farms 20 ha, but in common with most

Review

4 Explain the links shown in **7.1**.

5 Examine the view that women are the key players in the move to greater food security.

6 What do you think is meant by the empowerment of women? Give some examples.

small farmers she does not hold the title deeds of her land. It is ancestral land held by a local chief, but she is permitted to keep the holding so long as she or her children are capable of farming it. She used to grow just enough maize to feed her family, but in 1984 she took out a loan that allowed her to improve her farming methods. She also diversified her output to make better use of the rains, growing cabbages and other vegetables as well as maize. She now also has a small orchard of pear, peach, mango and mulberry trees. She keeps chickens and a few cattle, but the little meat eaten by the family is usually bought.

Thus Joyce Kayaya and her family enjoy a reasonable degree of food security in a country with serious economic problems. She is certainly not rich, but she is reasonably content with her lot. Clearly, her situation is so much better than that of Maria Mathai, who happened to be born in a country that still fails to recognise the vital role played by women in food production.

SECTION C

Dealing with land

The fact that land rights in most LEDCs do not extend to women is only one aspect of what might be called the 'land problem'. Two other vital issues are the amount of land available per family and the degree to which that land is secure in the hands of the family. On the latter, experience shows that improved farming and raised output are most likely where the tenure of the land is believed to be secure. This does not necessarily mean that every family should own the land that it farms, but at least the tenancy should guarantee their right to occupy and work it. As for the amount of land, one of the effects of rapid population growth has been the shrinking size of family farmsteads. In most LEDCs, the size of small family farms has been halved over the past four decades, as plots are divided into smaller and smaller pieces for each new generation of male heirs. In 57 LEDCs surveyed by the FAO in the 1990s, over half of all farms were were found to be less than one hectare in size. This is not large enough to feed a family with from four to six children. In India, 60 per cent of farms are less than one hectare in size.

At the same time, control over farmland is increasingly concentrated in the hands of large-scale farmers. In Guatemala, for example, 3 per cent of the farmers own 65 per cent of the most productive farmland. In Argentina, the pampas is divided into huge estates owned by a few families. Until recently, the 100 wealthiest families owned more than 4 million ha. In these and some other countries, production of export crops has taken priority over food crops, a policy that favours large-scale commercial agriculture. It also makes rural people more vulnerable to shifts in markets and fluctuations in commodity prices. Clearly, any suggestion that these large commercial farms might be broken up and their land redistributed to peasant families is bound to encounter immensely powerful opposition. Few governments would have the stomach for it, if only because such an action would prompt the loss of much-needed foreign currency, as well as the support of wealthy and influential people.

In theory, there is one other potential way of increasing the allocation of land per family, and that is to take more land into cultivation. The FAO has estimated that the present cultivated area in LEDCs might be increased by up to 40 per cent. But such potential farmland is highly marginal, with poor soils and either too little or too much rainfall. Bringing it into cultivation would require costly irrigation or drainage systems and large-scale soil fertility improvement programmes. At the moment, because of the technology and capital investment required, it looks as if any increases in food production will have to come from existing agricultural land.

Case study: Beating hunger in Brazil

There are estimated to be 30 million chronically hungry Brazilians. The government appears to be reluctant to give the eradication of hunger a high priority. Instead, it is being left to anti-hunger organisations and movements to take the initiative. They have already demonstrated that hunger may be overcome in a relatively short period of time, with simple actions and using existing technical, material and human resources. For example, a school feeding programme is reaching 30 million hungry Brazilian children daily. But the bonus of the programme is that it purchases food from local producers, thus providing those farmers with much needed capital to reinvest in their ventures.

In many parts of the country, where landless people have been given land, the transformation has been quite remarkable. It has been demonstrated that people entrusted with land under reform programmes can raise an annual income that is nearly four times the minimum annual wage. Landless labourers have been found to earn only three-quarters of that minimum wage. In those families that have been given land, infant mortality has fallen to half the national average. The lesson here is that land reform can create a small-farm economy that is good for local economic development. Its benefits are also social, in that poor people are no longer driven out of rural areas into towns and cities, where the chances of work are slim and living conditions are grim.

Case study: Playing the land reform card in Zimbabwe

Over the last few years, serious moves have been made by the government of Zimbabwe against White farmers, descendants of settlers who arrived here when the country was a British colony. The White farmers are responsible for growing a number of export crops, but principally tobacco. Tobacco and beverages account for roughly 30 per cent of all exports, and food and live animals another 11 per cent. The Black people in rural areas are either employed as landless labourers on the White farms or work small subsistence farms; some combine both.

Since 1980, when Zimbabwe gained its independence and Black majority rule began, the economy of this once prosperous part of Africa has gone into serious decline. Government corruption and misrule have been major contributors. Anxious to hold on to power, President Robert Mugabe and his political party have made land reform a major political issue, in the hope of using it as a smokescreen with which to hide their own disastrous management of the country, particularly the economy. In blaming the Whites for Zimbabwe's misfortunes, Mugabe has so whipped up feeling that many Black people have already forcibly occupied White-owned farms. This was in advance of legislation being passed through the Zimbabwe parliament to make the requisition of such farms 'legal'.

The outcome of this unhappy situation seems all too clear. The economy will plunge even deeper into debt and many more Black people will become responsible for raising their own food. The agricultural economy will become increasingly concerned with subsistence. The best that can be hoped is that that the new 'owners' of this redistributed farmland will show the enterprise needed to raise output and their own living standards.

Review

7 Why do you think that ownership of land is good for food production?

8 Evaluate what the redistribution of White-owned farmland will do for food security in Zimbabwe.

9 Explain why most governments tend to fight shy of land reform.

SECTION D

Cultivating water

When it comes to inputs into the food production systems of the hungry parts of the world, it might be argued that water is as important as land. Water has a double significance in the present discussion. It is needed for irrigation where rainfall is in short supply. It is also a medium for raising and harvesting food.

Although only 17 per cent of all crop land is under irrigation, this land produces over one-third of the global food supply. However, due to badly planned and poorly built irrigation systems, the yields on one half of all irrigated land have fallen in recent years. There are two main reasons for this: salinisation and the waterlogging of crops. In arid areas, salts that naturally occur in the soil must be flushed out with irrigation run-off. If not, they accumulate in the soil, eventually working their way to the surface, killing crops and poisoning the land. By the same token, improperly drained irrigation water can also raise the water table until it reaches the root zone and drowns the crop. The FAO estimates that salinsation has severely damaged around 12 per cent of the world's irrigated land, and that another 30 per cent has been adversely affected by a combination of salinisation and waterlogging.

With the spread of desertification, it is clear that there is going to be no let-up in the demand for irrigation. Clearly, the challenge here is to reduce the twin scourges of salinisation and waterlogging. Given the great scarcity of water resources, something also needs to be done to improve the general efficiency of irrigation. Currently, no more than half the water drawn for irrigation actually reaches the crops. The rest is lost as it makes its way to the fields, through evaporation, by soaking into unlined canals and through leakage from pipes. At present, 31 countries are experiencing chronic water shortages. That figure is expected to rise to 50 countries within the next 25 years.

The seas, lakes, ponds and rivers are traditional sources of food, mainly in the form of fish and crustaceans; human consumption of aquatic weeds continues in some parts of the world. Fish contributes a signficant amount of animal protein to the diets of people worldwide. It is highly nutritious and serves as a valuable supplement in diets that lack essential vitamins and minerals. As well as being an important source of food, it is also a source of work and money for millions of people. At least 200 million workers are dependent on fishing for their livelihood.

Figure 7.2 Integrated aquaculture in a Chinese village

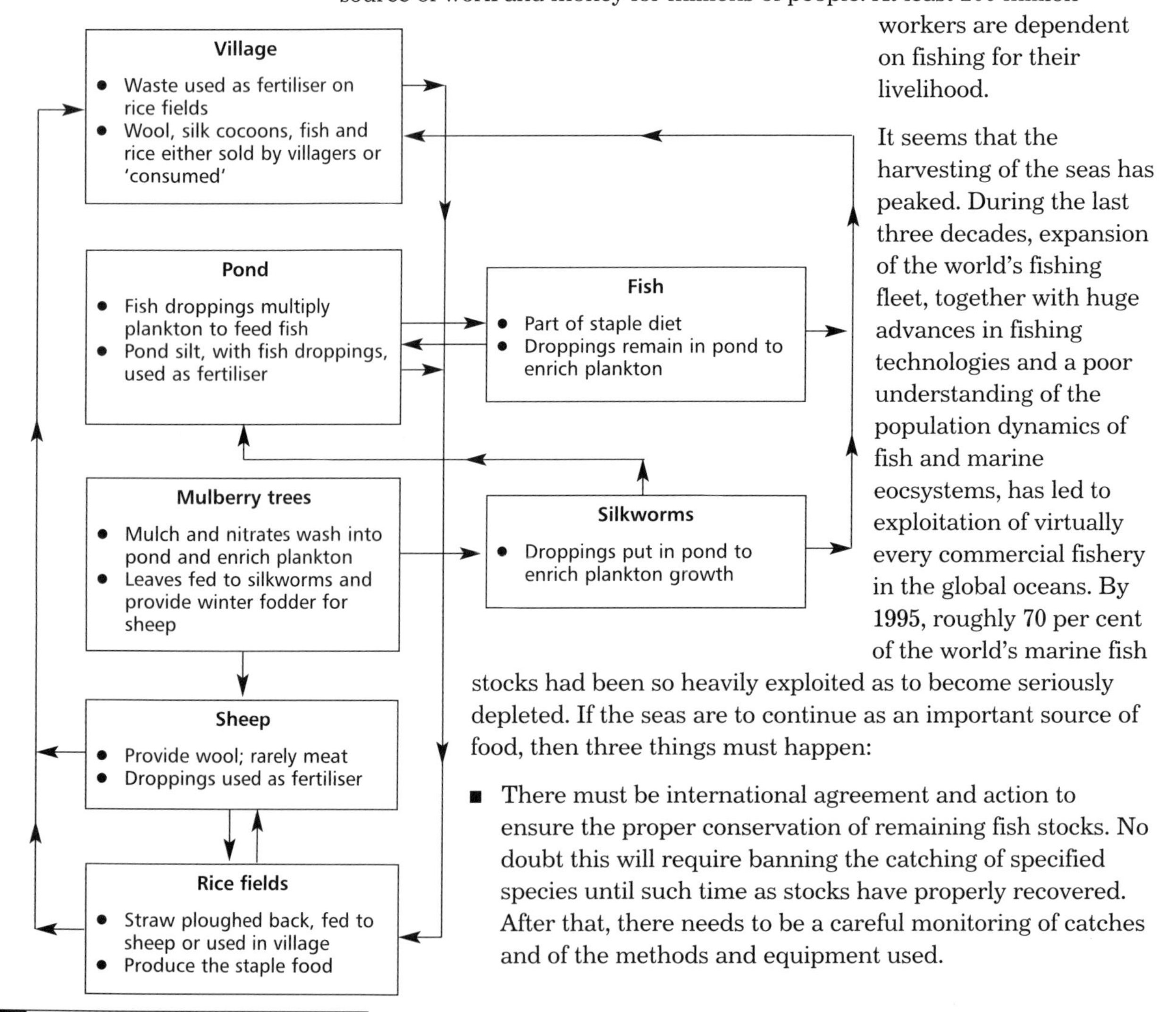

It seems that the harvesting of the seas has peaked. During the last three decades, expansion of the world's fishing fleet, together with huge advances in fishing technologies and a poor understanding of the population dynamics of fish and marine eocsystems, has led to exploitation of virtually every commercial fishery in the global oceans. By 1995, roughly 70 per cent of the world's marine fish stocks had been so heavily exploited as to become seriously depleted. If the seas are to continue as an important source of food, then three things must happen:

- There must be international agreement and action to ensure the proper conservation of remaining fish stocks. No doubt this will require banning the catching of specified species until such time as stocks have properly recovered. After that, there needs to be a careful monitoring of catches and of the methods and equipment used.

- The whole of the global fish catch must go for human consumption. At present, one-third of it is converted into fertilisers and animal feeds
- Fish-farming in inshore waters must be increased. Aquaculture has been growing fast enough to compensate for the decline in wild fish catches, but there have been pollution and contamination problems. Disaster may be just around the corner.

Aquaculture in freshwater environments has a long history. It is a line of food production that seems to offer some promise, at least in those areas free of water shortages. In China, fish-farming has been combined with other activities in an efficient and self-contained agricultural system (**7.2**). All of the components are interrelated, with waste from one becoming food or fertiliser for another.

Review

10 Is more irrigation the obvious way ahead in a world of water shortages?

11 Find out more about the pollution and contamination problems associated with marine fish-farming. You might focus on salmon farming.

12 Write a critique of the agricultural system shown in **7.2**.

SECTION E

Searching for sustainability

Sustainability is something of a buzzword these days, but it does have a particular relevance so far as agriculture and food supply are concerned. The decline in fish stocks discussed in the previous section is damning evidence of an unsustainable exploitation of the global seas. Recent approaches to agricultural development of the type discussed in **Chapters 4**, **5** and **6** have largely failed to reduce the absolute numbers of food-insecure people or to ensure minimal impact on the environment. While global achievements in raising food production have been impressive during the last 50 years, global disparities in people's access to food remain the biggest obstacles to achieving food security for all. It is a sad fact that many governments fail to recognise that there is much more to food security than simply producing more food, particularly in the risk-prone environments of LEDCs. This is where more sustainable forms of agriculture have much to contribute.

The sustainable production of food is the first requirement of food-secure livelihoods. Definitions are difficult, because one of the essential characteristics of sustainable agriculture is that it should be flexible and in tune with particular local conditions. Sustainable agriculture does not require the adoption of a predetermined package of technologies and practices. It is not a basic model that is imposed regardless. Rather, it is a process of learning and understanding, adapting and moving carefully forward. Nonetheless, it is possible to identify some underlying principles and characteristics:

- it emphasises management rather than technology
- it requires an understanding of natural processes and biological relationships
- it maintains biodiversity rather than reducing or simplifying it
- it minimises dependence on external inputs such as fertilisers and pesticides
- it maximises the regeneration of internal resources
- it encourages the participation of farmers and rural people in the processes of problem-solving
- it seeks to achieve equitable access to productive resources within local communities
- it encourages better use of local knowledge, local practices and local resources
- it aims to increase self-reliance among farmers and rural organisations.

Clearly, there is much listed here that needs to be explained and illustrated. This is perhaps best done by the following case study of permaculture.

Case study: Permaculture

In parts of the world where it is vital to increase food output – and to sustain that increase – it makes sense to use what the natural world provides in the best possible ways. Permaculture (permanent agriculture) does just that. The idea is a relatively new one. It was developed in Australia in the late 1970s by Bill Mollison. During the 1990s, the idea

Figure 7.3 A permaculture garden

really began to take off. Permaculture is knowledge-intensive, so training is needed. There are now over 100 permaculture training institutes in LEDCs.

In permaculture, farmers use no inputs, such as chemicals and pesticides, from outside the area where they farm. They grow a mixture of food and tree crops, and often keep a small number of livestock, with each part of the system benefiting from the other parts. Trees tap soil moisture, while leaf-fall from trees enriches the soil and helps crop growth.

Farmers who switch to permaculture may experience a temporary dip in yields, for it can take up to five years to wean the land off chemicals. Trees are planted around the fields and these soon produce mulch and nitrogen – natural fertilisers that are washed into the fields by rain. Often, fruit and vegetables are grown amongst the trees. Indigenous trees are preferred, because part of this farming technique involves making insecticide sprays from their leaves, bark and wood. All of the trees and crops are in a kind of symbiotic relationship, the idea being to balance sunlight and shade, pests and predators. The soil is never left exposed to the sun and wind; it is heavily mulched to keep it cool and damp. As one converted farmer has put it, 'All plants, insects, animals and we who tend the garden, live in a natural harmony.'

Experience has shown that permaculture can be adapted to work well in both humid and arid environments. In the latter, training courses provide instruction in simple but effective water conservation measures. In most cases, farmers can expect to increase their yields from upwards of four times within a few years. Besides delivering food security, the system allows farmers to reduce the proportion of their land used for immediate subsistence needs. Thus they have an opportunity to raise a cash crop or two.

As yet, governments have been slow to recognise the enormous benefits of permaculture. The governments of Thailand and Vietnam are the exceptions so far, but Bill Mollison does have flourishing training institutes in countries as far apart as India and Botswana. It is possible that governments are disinterested because they have a vested interest in other modes of farming that are largely geared to the unsustainable output of export crops. Perhaps they need to be informed that 'Adopting sustainable agriculture does not mean a return to some form of low-technology, "backward" or "traditional" agriculture.' It blends innovations that may originate with scientists, with farmers or with both. The local or grass roots character of permaculture and other sustainable forms of agriculture is critical, for this is the only way of ever achieving the triple aims of:

- reaching a level of output required for food security
- minimising damage to, and disturbance of, the local environment
- ensuring that future generations will continue to enjoy the same food security.

Review

13 Having read the permaculture case study, check it off against the principles and characteristics of sustainable agriculture listed just before it. How well does permaculture score?

14 Why is it that governments seem to be more interested in cash-cropping for export than in sustainable subsistence farming?

SECTION F

Conclusion

That concludes our look at some of the ways in which the food security of nations might be improved. Using a simple sixfold scheme, **7.4** classifies each action considered in this and the previous chapter according to:

- its potential impact
- its expected impact.

For example, aid has considerable potential to improve food security, provided that it is given in the right form. However, since the chances of that fully happening are rather remote, its actual impact is expected to be only slight. Readily apparent in **7.4** is the high scoring of the expected outcomes of those actions taken at a national level. This is encouraging, for it indicates that there is much that individual LEDCs can do for themselves – that is, if their governments have the will to do so, and to do so almost in isolation. There **is** a future for them without economic globalisation. There are opportunities within their grasp that need to be taken now – and rather than waiting, perhaps in vain, for MEDCs to change their rather exploitive and domineering behaviour. Hunger may be a huge and widespread problem, but it does not necessarily take global actions to conquer it.

For MEDCs, there are few worries about food security, except where there is a heavy dependence on imported food and therefore a reliance on other activities to yield the required revenue to buy that food. Possible outbreaks of disease remain a lurking fear, despite the advances of modern agriculture and medical science. There are two agricultural challenges in the North. One is to ensure that the dometic system produces food that is safe to eat and whose production is kind to the environment. The BSE scandal, the nitrates problem and the recent outbreak of 'foot and mouth'

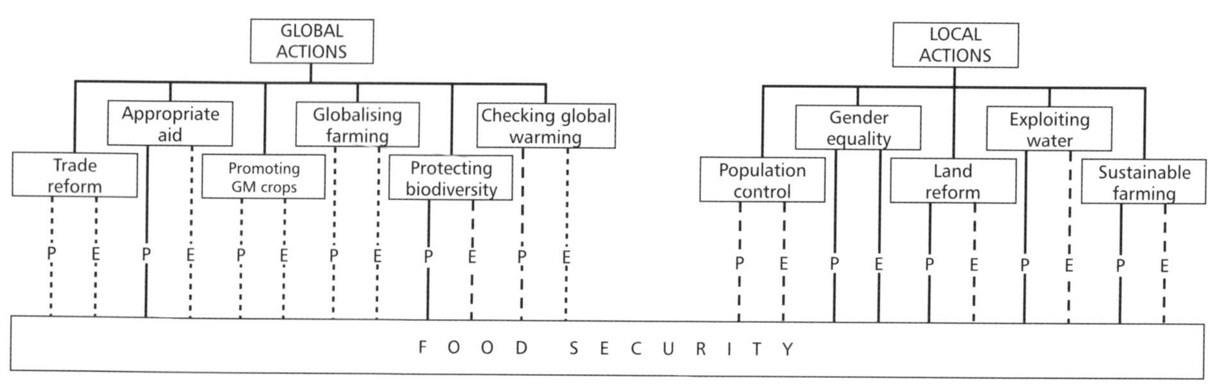

Figure 7.4 The potential and expected impacts of actions to improve food security

Key
P = potential E = expected
Positive impact:
——— considerable
– – – – moderate
- - - - - - - slight

disease are salutory reminders here. The second challenge is to reduce surplus production. Inevitably, this will involve taking land out of farming altogether. That, in turn, requires finding alternative means of livelihood to keep people in rural areas and maintain viable rural communities.

At the global level, if MEDCs are not prepared to promote genuine free trade and forego their urge to 'rule' the world through TNCs, trading blocs, tied aid and other globalisation processes, then so be it. One can only regret such unwillingness to work for the global good. They could do so much to help foster farming, further food production and fight famine in less fortunate parts of the world. If the North was to do only one thing to help the South, then best by far would be to cancel the crippling debt burden. With the slate wiped clean, LEDCs would be in a much better position to begin implementing and coordinating the sorts of national and local initiatives evaluated in this chapter. As such, this would mark the start of a path leading towards a world in which no one is excluded because of a lack of food.

Review

15 To what extent do you agree with the impact assessments that have been made in **7.4**? Where you strongly disagree, give your reasons.

16 What conclusions do you draw about the relative impacts (potential and expected) on food security of global and local actions?

17 To what extent do you agree with the conclusion that better food security is more likely to come from local rather than global actions?

Enquiry

1 The following are other actions that might be taken to improve food security:

- water conservation
- intercropping
- urban farming.

Select **one** and research it, paying particular attention to:

a what it involves
b how feasible it is
c its potential and expected impacts on food security.

It might be appropriate to anchor your investigation to one or two LEDCs.

2 Investigate the case for and against the UK relying less on imported food.

Further reading and resources

There are frequent articles providing information about farming and food issues in periodicals such as *The New Internationalist, People and the Planet* and *The Ecologist.*

The internet can be extremely helpful. Many organisations related to farming and food have their own websites, as do the major TNCs, food manufacturers and food retailers. The following list summarises the websites mentioned in the text:

Action Aid: http://www.actionaid.org/home.html
AstraZeneca: http://www.astrazeneca.com/
Aventis: http://www.aventis.com/
Cafod: http://www.cafod.org.uk/
DuPont: http://www.dupont.com/
FAO: http://www.fao.org/
Farm Africa: http://www.farmafrica.org.uk/
Intermediate Technology Development Group: http://www.itdg.org/
Nestlé: http://www.nestle.co.uk
Novartis: http://www.novartis.com
Tesco: http://www.tesco.com
Unilever: http://www.unilever.com
Water Aid: http://www.wateraid.org.uk/

Some book references include the following:

Ian Bowler, *Agricultural Change in Developed Countries* (Cambridge University Press, 1996).

John Madeley, *Hungry for Trade* (Zed Books, 2000).

John T. Pierce, *The Food Resource* (Longman, 1990).

Geoff Tansey and Tony Worsley, *The Food System: a Guide* (Earthscan, 1995).

Elizabeth M. Young, *World Hunger* (Routledge, 1997).